EINSTEIN, HISTORY, AND OTHER PASSIONS

Other Books by Gerald Holton

INTRODUCTION TO CONCEPTS AND THEORIES IN PHYSICAL SCIENCE, 1952, 1973, 1985.

FOUNDATIONS OF MODERN PHYSICAL SCIENCE with D. H. D. Roller, 1958.

THEMATIC ORIGINS OF SCIENTIFIC THOUGHT: KEPLER TO EINSTEIN, 1973, 1988.

THE PROJECT PHYSICS COURSE, with F. J. Rutherford, F. Watson, and others, 1970 and later editions.

SCIENTIFIC IMAGINATION: CASE STUDIES, 1978.

THE ADVANCEMENT OF SCIENCE, AND ITS BURDENS: THE JEFFERSON LECTURE AND OTHER ESSAYS, 1986.

SCIENCE AND ANTI-SCIENCE, 1993.

GENDER DIFFERENCES IN SCIENCE CAREERS: THE PROJECT ACCESS STUDY, with Gerhard Sonnert, 1995.

WHO SUCCEEDS IN SCIENCE? THE GENDER DIMENSION, with Gerhard Sonnert, 1995.

Books Edited by Gerald Holton

SCIENCE AND THE MODERN MIND, 1958.

EXCELLENCE AND LEADERSHIP IN A DEMOCRACY, with S. R. Graubard, 1962.

CLASSICS OF SCIENCE, 1963–76.

P. W. BRIDGMAN, COLLECTED EXPERIMENTAL PAPERS (member of editorial board), 1964.

SCIENCE AND CULTURE: A STUDY OF COHESIVE AND DISJUNCTIVE FORCES, 1965.

THE TWENTIETH-CENTURY SCIENCES: STUDIES IN THE BIOGRAPHY OF IDEAS, 1972.

SCIENCE AND ITS PUBLIC: THE CHANGING RELATIONSHIP, with W. A. Blanpied, 1976.

LIMITS OF SCIENTIFIC INQUIRY, with Robert Morison, 1979.

ALBERT EINSTEIN: HISTORICAL AND CULTURAL PERSPECTIVES, with Y. Elkana, 1982.

HISTORY OF MODERN PHYSICS AND ASTRONOMY, with K. R. Sopka, 1983–.

VISIONS OF THE APOCALYPSE: END OR REBIRTH?, with S. Friedlander et al., 1985.

COLLECTED PAPERS OF ALBERT EINSTEIN (member of editorial board and editorial committee), 1987–.

EINSTEIN, HISTORY, AND OTHER PASSIONS

*The Rebellion against Science at the
End of the Twentieth Century*

GERALD HOLTON

ADDISON-WESLEY PUBLISHING COMPANY

*Reading, Massachusetts   Menlo Park, California   New York
Don Mills, Ontario   Harlow, England   Amsterdam   Bonn
Sydney   Singapore   Tokyo   Madrid   San Juan
Paris   Seoul   Milan   Mexico City   Taipei*

Library of Congress Cataloging-in-Publication Data
Holton, Gerald James.
Einstein, history, and other passions : the rebellion against science
at the end of the twentieth century / Gerald Holton.
p.   cm.
Originally published: Woodbury, NY : AIP Press, © 1995.
Includes bibliographical references and index.
ISBN 0-201-40716-7 (alk. paper)
1. Science—Miscellanea.   2. Einstein, Albert, 1879–1955—
Knowledge—Science.   3. Science—Study and teaching.   I. Title.
Q173.H7342   1996
306.4'5'0904—dc20   95-50608
                                                                      CIP

An earlier version of this book was published by the American Institute of Physics Press in 1995.

Cover design by Suzanne Heiser
Text design by Irving Perkins Associates
Set in 11-point Janson by Pagesetters, Inc.

1 2 3 4 5 6 7 8 9-MA-0099989796
First printing, April 1996

Dedicated to the innumerable men and women in the twentieth century who devoted their lives to the advancement of science and the betterment of the human condition—who believed, with Thomas Jefferson, that "knowledge is power, knowledge is safety, knowledge is happiness."

CONTENTS

PREFACE

ONE OF THE burning questions of the day concerns the rightful place of science in our culture. The main purpose of this book is to make this debate more understandable—first by baring its historic roots and then by focusing, as a concrete example, on Albert Einstein's profound and lasting impact on our civilization. What can we learn about the powers and limitations of science by tracing Einstein's way of thinking, his view of the world, even his personal life?

Ours is an especially opportune time to examine these questions. In every society, the way in which science—as a body of knowledge, as a source of technical applications, as a generator of models for thinking and acting, as a troubling challenger of established ideas—is viewed and used affects its moral authority, much like the other significant components of a culture, such as religion and art. Indeed, in C. P. Snow's sharp formulation, science and its applications can determine human destiny, "that is, whether we live or die." It has been so throughout history. Sultan Muhammad II used technical innovations to pound Byzantium into submission in 1453; an Asian emperor in the nineteenth century made the fateful error of isolating his people from Western knowledge and might; in World War II, the perfection of radar provided a crucial edge for warding off the fascists' challenge to Western civilization itself. Equally essential has been the role of science and technology in mankind's ancient wars against ignorance, disease, and other blights and human burdens. Even the notion of human rights has been expanded by scientific findings; thus, modern anthropologists have exposed the falsity of the centuries-old idea about "inferior races," on which bigotry has long rested comfortably; and biomedical advances in contraception have had a liberating effect on women around the world.

But every age rethinks what its culture is and should be, what roles its components play. And as in many periods in the past, we are today again in the middle of a serious debate. Some academics, students, statesmen, policy makers, religious leaders, and other citizens— embracing a variety of ideological positions, and some of them frightened by abuses of science and technology as the result of corporate or governmental policies—are challenging the very legitimacy of science and technology in our culture and social fabric. For a decade or so, among influential intellectuals outside science, and gradually among segments of the general population, a seismic shift has occurred in the belief, espoused since the Enlightenment, that science and technology are on balance predominantly positive forces.

Ironically, this counter-movement is asserting itself just when the understanding of natural phenomena, the methods for reaching such understanding, and the agreement among scientists on responsible conduct, are all at their highest point. Yet the notion of progress, outside the laboratory, is considered by many "an idea whose time has passed," to quote the call to a recent conference at one of America's most prestigious universities. The term *neo-Luddite* is becoming a badge which some wear proudly. As Isaiah Berlin noted in his most recent book, no one predicted that the current version of the recurring historic phenomenon, called the Romantic Rebellion, against such notions as rationality and objectivity, would attain dominance in the last third of the twentieth century, that we would see again "the rejection of reason and order as being prison houses of the spirit."

In this atmosphere, it has become easy to overturn a long-standing social contract. President Roosevelt, at the end of World War II, received a report from a group of scientists, engineers, and other intellectuals (the so-called Vannevar Bush report), which promised "a fuller and more productive life" if scientific research were allowed to flourish. That set the stage for major financial and political support of science. Now, major industrial laboratories and the new majority in the United States Congress are drastically cutting back spending for civilian research and development, even though economists have shown that the annual rate of social return on investments in research and development is remarkably high. Similarly, while the authority of scientific thought has been decreasing—and the scientific community so far has been on the whole ineffective in reevaluating and reasserting its sense of self—an ever-widening gap separates the corpus of knowledge and the scientific worldview on the one hand and the popular understanding of these on the other.

In some of the foremost universities in the United States, the attention to science and mathematics required from college students in the total curriculum ranges from zero to a mere 6 percent. Into this vacuum of scientific illiteracy among our future leaders are rushing bizarre notions about science, scientists, and their roles in society. These are eloquently propagated by ambitious factions with a wide range of motivations, from the ideological to the supernatural. The spokespeople for this movement include, for example, well-placed academics and New Age fanatics; museum curators who were reported to be intent on showing that science equals "pollution and death"; a widely read radical who claims that Newton's *Principles of Mechanics* is merely a "rape manual"; and a social scientist who announced, "There is no Nature," only "a communication network" among scientists. In the present climate, such depictions of science have become more and more prominent in the marketplace of ideas, going far beyond the reasonable and necessary scrutiny of abuses and limitations of this, as of any, human enterprise.

The early chapters of this book provide the historical context and taxonomical order of the present controversy over the rightful place of science in our society. They set the stage for considering the interactions—some plainly visible, others subterranean but equally vital—between the world of science and the other components of our culture. What are the major images of science among the public? What are the signs of dangerous excess, both of uncritical "scientism" on one side and of antiscience on the other? How did trust in scientific findings evolve, and what are its limits? How indeed does the individual scientific imagination work, including both reason and intuition, logic and thematic choice? What does it take to understand an event in the history of science?

These matters, discussed in Part One, are followed in Part Two by exemplifications from the life and work of a scientist who had a transforming effect on our era. There I trace in specific terms the influence of Albert Einstein on the culture of his century. His carefully thought-through considerations of the meaning of progress and goals for science remain fully applicable today. Through his personal papers we can discern how, as a young scientist, he tried to build a life in which work and love were made into a seamless whole; we can follow his analysis of how to think effectively, not only about scientific problems; and we shall see in him an example of the creative role of a rebel operating *within* the scientific tradition.

My foremost hope is that this volume will convey the civilizing power of scientific thought, and thereby enable the reader to participate more actively and confidently in today's cultural debate.

Part One

SCIENCE IN HISTORY

Chapter 1

WHAT PLACE FOR SCIENCE AT THE "END OF THE MODERN ERA"?

For about a decade, a movement among a segment of academics, eloquent popularizers, and policy makers has been mounting a challenge to the very legitimacy of science in our culture. Through these efforts, concepts such as the "end of the modern era," the "end of progress," and the "end of objectivity" are making an unquestioned place for themselves in the public mind. Far from being a passing phase or a *fin-de-siècle* preoccupation, this movement signals the resurgence of a recurring rebellion against some of the presuppositions of Western civilization derived from the Enlightenment period. A chief object of this countercultural swing is to deny the claim of science that it can lead to a knowledge that is progressively improvable, in principle universally accessible, based on rational thought, and potentially valuable for society at large. The impact of this reviving rebellion on the life of the scientist, on the education of the young, on public understanding of science generally, and on the legislation of science support is measurably growing, and has become visible even to the least attentive.

The aim of these chapters is to study this movement, and its driving ambitions, in the hope of understanding it. To do so we must first consider the views of some of the twentieth century's chief theorists of science and culture, which laid the foundation for the implicit "contract" forged in the aftermath of War War II between science and society.

That contract, still the dominant image among the majority of scientists even while it hardly corresponds to reality today, was the result of a more innocent phase. For a few decades the pursuit of scientific knowledge was widely thought—above all by scientists themselves—to embody the classical values of Western civilization, starting with the three primary virtues of truth, goodness, and beauty. That is, science tended to be praised as a central truth-seeking and enlightening process in modern culture, what one might call the Newtonian search for omniscience. Science and scientists were also thought to embody the ethos of practical goodness in an imperfect world, both through the largely self-correcting practice of honor in scientific research, and through applications that might improve the human condition and ward off the enemies of society, a sort of Baconian search for a benign omnipotence. Finally, science was also thought of as a Keplerian enchantment; the discovery of beauty in the structure, coherence, simplicity, and rationality of the world was the highest reward for the exhausting labor the discipline required.

BEFORE THE EUPHORIA ENDED

The last time the optimistic portrayal given above could have been said to be generally taken for granted, at least in the United States, was the period following the end of World War II. It was embodied also in the famous Vannevar Bush report, *Science, the Endless Frontier*, of 1945, which became a driving force of U.S. science policy. Because it is such a convenient example of modern post-Enlightenment optimism about the role of science in culture, one which so many scientists tacitly assume to be still operative, it will be illuminating to look at the main thrust of that document.

In November 1944, President Roosevelt requested from Vannevar Bush, the head of the wartime Office of Scientific Research and Development, a report that would outline how, in the postwar world, research in the natural sciences—he called it "the new frontiers of the mind"—could be strengthened and put into service for the nation and humanity. Roosevelt was particularly interested in three results: waging a new "war of science against disease," "discovering and developing scientific talent in American youth," and designing a new system of vigorous federal support for scientific research in the public and private sectors. Above and beyond those, he argued that science's applications, so useful during the bitter war to preserve the world from fascist dictatorship (with the successes of the Allies' radar, antisubmarine devices, and synthetic rubber the most striking examples at that time), could now be harnessed to "create a fuller and more fruitful employment, and a fuller and more fruitful life."

Vannevar Bush's detailed response came less than eight months later, the result of a crash program by an impressive brain trust of about forty experts from industry, academe, and government. Roosevelt had died, but with the war's successful end in sight, the American administration proved generally hospitable to the report's ideas. While some of the details were overly optimistic and others were modified in practice (often to Bush's dismay), his vision, it is generally agreed, set the stage for the development of new institutions for the support of science during the following decades, and paralleled the generally favorable popular attitudes that were prerequisites for action. The base was laid for global leadership in many branches of basic science. Not until the Vietnam War escalated was there substantial popular disenchantment both with governmental authority and with the widely visible use of sophisticated technology in a hopeless and unpopular war, and by implication with science that presumably could help give birth to such abuse. This turn signaled the end of what might be called a rather euphoric phase in the relation of science and society in this century.

The Bush report, as well as the rival proposals by Senator Harley Kilgore, were historic exemplars of the science-based progressivism in post-World War II America, which saw science and democracy as natural allies in the service of the ideal of empowerment and instruction of the polity as a whole. In this sense, they were part of the American dream as far back as Benjamin Franklin and his fellow statesmen and science amateurs. Vannevar Bush himself hinted as much in the brief preface to his report, taking courage from the fact that, as he put it, "the pioneer spirit is still vigorous within the nation." And to make the connection with the tradition of Condorcet even more explicit, he added a sentence which, while presenting the reigning opinion of a citizen of the mid-1940s, is likely to be rejected today by many who think of themselves as the children of the 1960s and 1970s. He wrote: "Scientific progress is one essential key to our security as a nation, to our better health, to more jobs, to a higher standard of living, and to our cultural progress." One could hear an echo of Thomas Jefferson's formula: "The important truths [are] that knowledge is power, knowledge is safety, knowledge is happiness."

Bush and his contemporaries could hardly have imagined that by the early 1990s those hopes would have begun to be rejected, even at the highest levels—that, for example, a key person in the U.S. Congress for science policy would imply (as we shall see in more detail later) that science and technology alone can be held to account for the whole sorry list of failures of decades of misdirected political and business leadership; he said: "Global leadership in science and technology has not translated into leadership in infant health, life expectancy, rates of literacy, equality of opportunity, productivity of workers, or efficiency of resource consumption. Neither has it overcome failing education systems, decaying cities, environmental degradation, unaffordable health care, and the largest national debt in history."[1] And another highly placed observer, formerly the director of the National Science Foundation, exulted: "The days of Vannevar Bush are over and gone . . . the whole world is changing."

THE CHANGING BALANCE OF SENTIMENTS

We turn now from momentary vagaries to a look at the causal mechanisms responsible for the changes in the place assigned to science at significant stages in the intellectual history of the past hundred years. For if we know the general causes in the variation of the underlying ideology, we shall better understand the changes in policy toward science at any given moment.

Here we must confront at once the question of whether these changes are gradual, and part of an evolutionary development, or are so sudden that, as if in a political revolution, one passes discontinuously from the end of one age to the beginning of another. If the latter is the case, we would now be passing through a rupture of history, with "modern" behind us and "postmodern" right, left, and all before us. While I doubt this is the case—and it certainly is not visible in the *content* of science as against some of the current writings about science today—a fashion in history proper has for some time been the attempt to discern the arrival of a new age. Periodization, the arranging of the flow of events into clearly separate eras, is a common tool, although applied more wisely from the safe distance of retrospection. That is how we got such schoolbook chapters as "The Age of Reason" and "The Progressive Era in America" around the turn of the century.

A chastening example of periodization was provided by the American historian Henry Adams. At the beginning of the twentieth century, he had been impressed by the publications of the chemist J. Willard Gibbs of Yale on the phase rule for understanding heterogeneous equilibria. Adams was also fascinated by the strange idea of some physicists of that day that the phase rule can serve, by analogy, as a means for putting into hierarchical order the following sequence: solid, fluid, gas, electricity, ether, and space, as if they formed a sequence of phases. Stimulated by such ideas, Adams believed that thought itself passed in time through different phases, each representing a different period. In his essay of 1909, "The Rule of Phase Applied to History," Adams came to a remarkable conclusion about the imminent passing of modernity: "The future of Thought," he wrote, "and therefore of History, lies in the hands of the physicist, and . . . the future historian must seek his education in the world of mathematical physics. . . . [If necessary] the physics departments will have to assume the task alone." Henry Adams's conclusion might fairly have been called in its own day a declaration of what the postmodern age would look like. But today's formulation is likely to state the exact opposite.

I cite this example—and many others come to mind—to signal my discomfort with the division of history into distinct periods. A less rigid and more workable notion is to recognize that at any given time and place, even during a period when a civilization appears to be in a more or less settled state of dynamic equilibrium, there exist simultaneously several competing and conflicting ideologies within the momentary, heterogeneous mixture of outlooks. As Leszek Kolakowski noted, "It is certain that modernity is as little modern as are the attacks on modernity. . . . The clash between the ancient and the modern is probably everlasting and we will never get rid of it, as it expresses the natural tension between structure and evolution, and this tension seems to be biologically rooted; it is, we may believe, an essential characteristic of life."[2]

It is sometimes possible in retrospect to identify one of the competing worldviews as the most dominant one for a longer or shorter period. But what is also likely to occur when followed in real time are two effects. The first is that each of the different competing groups works fervently to raise its own ideology to a position where it would be accepted as the "taste of the time" or the "climate of opinion" which characterizes that particular age and region. The newest and most ambitious one *will also be trying as part of its agenda to delegitimate the claims of its main rivals.* Especially when the previously relatively stable equilibrium begins to crumble, the pandemonium of contrasting voices gets louder. Some partial victors rise to be major claimants above the rest, and one of them may even be generally recognized for a while as the embodiment of the new worldview or "sentiment" of the society.

In addition, in this constant seesaw of changing historic forces, mankind's inherent liability to engage in overambition or one-sidedness may infect some of these claimants (not excluding, on occasion, scientists). This is the tendency, as Hegel had warned, toward "the self-infinitization of man," or simply to *yield to excess*—which in turn can generate the same sort of excess among the opposing claimants. Recognizing these two facts—the mutual attempts at delegitimation and the tendency to yield to excess—is central for understanding the course of cultural conflict, today as in the past.

In this general struggle, from that of Apollo vs. Dionysus in Greece to this day, the specific, more limited question of the place assigned to the scientific conception of the world has always played a part. Sometimes this place has been at the cherished core of the rising or victorious overall worldview; sometimes it has found itself embedded in the sinking or defeated one, and then was even accused of nourishing a great variety of sins against the better interests of humanity.

Historians of ideas have mapped the changing forms of the general contrary trends. Wise political leaders, too, have at times watched with apprehension as the net balance of prevailing sentiments has taken a turn, for as Jefferson said, "It is the manner and spirit of a people which preserve a republic in vigor. A degeneracy in these is a canker which soon eats into the heart of its laws and constitution." Weighty scholarship has chronicled how one of the world conceptions, and the scientific position within it, gained predominance over the others for some decades in significant segments of Western culture—an example is

Robert K. Merton's early study on science and seventeenth-century Puritanism. There is also much documentation of how such sentiments subsequently gave ground, as the overall balance of benignity or distress moved the other way. As to the practicing scientists themselves, whether for reasons of preoccupation or timidity, most of them have paid little attention to this constant seesaw of sentiments, except to weigh in now and then as promoters of the positive swings, or occasionally to become victims of the negative ones.

Today, at our own *fin de siècle*, this oscillating spectacle, so engrossing to the scholar, has ceased to be merely the site for research by historians. The general balance among the contending elements, and with it the attitude of traditional patrons, is changing before our eyes. Studying this current drama is as fascinating and fruitful for the historian of ideas, whose perspective I shall be taking here, as the appearance of a supernova may be for an astronomer. But in both cases, the present state is the product of a historic process, the latest member of a motley progression.

Toward a "Monistic Century"

Let us therefore look at some of the ideologies that have claimed to represent the climate over the past hundred years or so up to the present—a sequence of selected samples meant to be analogous to stages in the growth of a culture of cells seen under the microscope. Our first sample concerns an event that occurred as the new century was signaling its beginning: the World's Columbia Exposition at Chicago in 1893. The fair was intended as a triumphant celebration of human and social progress in all fields—above all, industrial, scientific, and architectural. The big attractions were Machinery Hall, the Electricity Building, the Electric Fountain, and the halls on Transportation and Mines. On the opening day, President Grover Cleveland was on hand to push a button that turned on an abundance of electric lights and motors. (Electric light bulbs and AC motors were still fairly new.) This caused such an excited forward surging of the thousands of spectators that many fainted in the crush. One may safely assume that few among the twenty-seven million attendees during the exposition worried about, say, the ill effects of rapid industrialization. And few if any would have guessed that, just a century later, at a World's Fair held in South Korea, the official U.S. exhibit, as if in obeisance to a new *Zeitgeist*, would be dedicated to the detritus of the postindustrial world, featuring mounds of broken machinery and views of festering nuclear disposal sites; or that the new permanent exhibition at the Smithsonian Institution's Museum of American History, "Science in American Life," would devote much of its space to an exposé of the hazards of science and the public's alleged disillusionment with technology.

Another indication of how much the worldview changed during one century is that one of the major events of the exposition of 1893 was a spectacular World's Parliament of Religions. Personal religion is, and always has been, close to the hearts of most Americans. But it now seems surprising that on that occasion, in a setting glorifying science and industry, hundreds of religious leaders from all parts of the world met to present their views in two hundred sessions during seventeen days. It was a colorful affair, with Hindus, Buddhists, Jains, Jews, Protestants, Catholics, adherents of Shinto and Zoroaster, and so forth, all meeting together in their robes "for a loving conference," in the words of the chairman of the parliament, J. H. Barrows. The purpose was clear. As it was for the exposition as a whole, the subtext of that Parliament of Religions was progress and harmonious unity. Hence the exposition, Barrows said, could exclude religion no more than it could exclude electricity. Science was invoked as an ally in reaching a higher unity while serving the needs of mankind.

One of the passionate believers that science, religion, and indeed all cultural activities are aspects of one grand unification program was one of the organizers of the Parliament of Religions, Paul Carus, a publisher now remembered mainly for having brought the writings of Ernst Mach to readers in the United States. The title of his presentation[3] was nothing less than "Science, a Religious Revelation." His was a sort of anticlerical post-Christian deism, much of which would have appealed to some American statesmen-philosophers of an earlier century. Individual dignity, Carus thought, can only be found through the discovery of truth, and that is the business of science. Hence, he announced, "Through science, God speaks to us." One did not have to choose between the Virgin and the Dynamo; rather, the laboratory was the true cathedral, and vice versa. As the masthead of his journal *The*

Open Court put it, he was "devoted to the science of religion [and] the religion of science. . . ."

Carus typified a popular, science-favoring universalism of that time which today is severely challenged, from the Right and from the Left. I have chosen Carus because his world picture is a good example of a movement then prominent: Modern Monism, based on the belief in a "unitary world conception." It arose essentially as an antithematic response against the Cartesian dualism of the material versus the mental, and against the multiplicity of commonsense experience, with its starting point in personal individuality. The movement on behalf of Monism had the enormous ambition, in the words of Carus, "to direct all efforts at reform, and to regenerate our entire spiritual life in all its various fields." This meant of course replacing conventional religion with what Carus called the "Religion of Truth," where Truth is defined as "the description of fact . . . ascertainable according to the methods of scientific inquiry." In this sense, "science *is* revelation"; and in this way one would overcome the old, unacceptable dualism of scientific truths versus religious truths.

The head of the small but ambitious international monistic movement was the great German chemist Wilhelm Ostwald (Nobel Prize, 1909). Whereas most modern scientists are quite aware of the limits even within their research—as Max Planck said in 1931, "A science is never in a position completely and exhaustively to solve the problem it has to face"—the publications of the monistic movement show that it hoped every aspect of culture, life, and society would be guided by monistic ideas, from the education of children to the economy of nations, and of course within the research program of science itself. Thus Ernst Haeckel, another patron of the movement, predicted that physical science would eventually trace back all matter to a "single original element."

Despite the philosophical naïveté of its leaders, the movement attracted for a time an enthusiastic following. In Germany, it had branches in forty-one cities and even organized public mass demonstrations against the Church. One must perhaps allow for the effects on them of having to live under the reactionary political clericalism of Germany. But I have intentionally chosen this case of "scientism," of excess on the part of a small minority of scientists, as my first example of *the rhetoric of a polarizing overreaching by many movements, before and since, on either side.* Thus, caught up in this fervor, Ostwald, with hubris unequaled by the few remaining captives of scientism today, was propelled to the heights of overambition, with such claims as these in 1911: "We expect from science the highest that mankind can produce and win on this earth. . . . Everything that mankind, in terms of its wishes and hopes, its aims and ideals, combines in the concept God, is fulfilled by science." And finally, "Science, now and with immeasurable success takes the place of the divine." Ostwald added the prophecy that "we see arrive the Monistic Century. . . . It will inaugurate a new epoch for humanity, just as 2,000 years ago the preaching of the general love for humanity had inaugurated an epoch."[4]

Only a year after this publication, neither the Monistic nor the Christian base for kindness and love of fellow man had triumphed. Instead, war, which William James called the inevitable "bloody nurse of history," had taken charge. Strangely enough, it was Henry Adams who had sensed that the trend would be ultimately against a monistic century. Writing in 1905 in his autobiography, *The Education of Henry Adams*, he identified the course of history as away from unity and toward fragmentation and multiplicity. Indeed, in the aftermath of World War I, the idea of progress, and optimism about the place of science in culture, were war casualties. The balance had swung the other way. The only major movement with large political ambition that continued to claim a scientific basis was of course Marxism, especially as defended by Lenin in his 1908 book, *Materialism and Empirio-Criticism*. The assertion that Marxism-Leninism, the founding ideology of the Soviet Union, had anything to do with real science, was a purely rhetorical device, one of this century's great delusions even if this propaganda was taught to every child in communist countries. It is disproved, not least by the faulty analysis of science and its philosophy in Lenin's own book, and by the widespread mistreatment which Soviet scientists experienced when their theories did not please their governments.

SPENGLER'S PREDICTION OF THE END OF SCIENCE

Perhaps the most widely read attack against the claims of science appeared as the war was ending in 1918, and later it deeply influenced such theoreticians of history as Arnold Toynbee and Lewis Mumford.

The book was called *The Decline of the West*, written by a German mathematics teacher, Oswald Spengler. No quick summary can do justice to that richly baroque work, but the point I want to focus on here is what it had to say about the topic before us. Spengler's key conception was that for every part of mankind, in every epoch since Egypt, Greece, and Rome, the history of a civilization has taken fundamentally the same course, and this will continue in the future. Thus our own inevitable destiny in the West is to go to dust according to a timetable that can be calculated from the available precedents. Spengler predicted the very date of our undoubted demise: the year 2000.

The end stages of every civilization, Spengler wrote, can be recognized by the ideas science treasures in its own progress—by the adoption of the notion of causality instead of destiny; by attention to abstractions, such as infinite space; and to cause and effect, rather than to "living nature." The primacy of the soul is replaced by intellect; mathematics pervades more and more activities; and nature is reinterpreted as a network of laws within the corpus of what Spengler calls "scientific irreligion." Here Spengler introduces his most startling idea, one that has become familiar in new garb also. He warns that it is characteristic of the winter phase of civilization that precisely when high science is most fruitful within its own sphere, the seeds of its undoing begin to sprout. This is so for two reasons: the authority of science fails both within and beyond its disciplinary limits; and an antithetical, self-destructive element arises inside the body of science itself that eventually devours it.

The failure of science's authority outside its laboratories, Spengler says, is due in good part to the tendency to overreach and misapply to the cosmos of history the techniques that are appropriate only to the cosmos of nature. Spengler holds that the thought style of scientific analysis, namely "reason and cognition," fails in areas where one actually needs the "habits of intuitive perception" of the sort he identifies with the Apollonian soul and the philosophy of Goethe. But asserting that an unbridgeable contrast exists between a pure "rationality" of abstract science and the intuitive life as lived, Spengler commits the same error as all such critics before him and after, of whom few seem even to have come closer to science than through their school textbooks. Therefore they are ignorant of the vast difference between, on the one hand, "public science"—the final results of intersubjective negotiations to fashion at least a temporary consensus on the basis of experiment and logic—and on the other hand, the earlier, "private" stage of work in science, where the particular researcher's intuitive, aesthetic, thematic, or other nonlogical preference may be the key to the individual's advance beyond the previous level of public science. The complementarity between these two quite different stages in the actual development of any scientific result explains why in any given field the findings by natural scientists, operating within vastly different cultures and styles, are eventually harnessed into common products with global validity.

All this may be clear enough to practicing scientists. But, Spengler continues, even in the cosmos of nature there is an attack on the authority of science, arising from within its own empire: Every conception is at bottom "anthropomorphic," and each culture incorporates this burden in the key conceptions and tests of its own science, which thereby become culturally conditioned illusions. All our rushing after positive scientific achievements in our century only hides the fact, he thinks, that as in classical times, science is once more destined to "fall on its own sword," and so will make way for a "second religiousness."

What Spengler termed the orgy of two centuries of exact sciences would shortly be ending, together with the other, more valuable components of valuable Western civilization. As a kind of postscript, Spengler added his opinion in his later book, *Man and Technics* (1931), that advancing technology, with its mindlessly proliferating products, will also turn out to undermine the society of the West—because, he prophesied, its interest in and support of science and engineering will decrease: the "metaphysically exhausted" West will not maintain advances in these fields. Instead, the previously overexploited races in the rest of the world, "having caught up with their instructors," will surpass them and "forge a weapon against the heart of the Faustian [Western] Civilization." The non-Caucasian nations will adopt the technical skills, excel in them, and turn them against the Caucasian originators. In short, as H. Stuart Hughes put it, Spengler's prediction was that the East will triumph through better technology, first in commerce, and then militarily.[5]

A "SCIENTIFIC WORLD CONCEPTION"—THE VIENNA CIRCLE

The early response to Spengler's diagnosis was predictably bimodal— on one side there was wide and enthusiastic acceptance, which continues among people today who have never read Spengler but, so to speak, have imbibed his ideas with their mother's milk. On the other side, the opponents of Spenglerian scenarios included of course many prominent scientists. Some of these had joined in a study group that called itself the Vienna Circle, which met in the 1920s and early 1930s for discussion and publication. It included Moritz Schlick, Rudolf Carnap, Philipp Frank, Kurt Gödel, and Otto Neurath. Among their active sympathizers, they could count Hans Reichenbach and Richard von Mises in Germany, and in America, B. F. Skinner, P. W. Bridgman, Charles Morris, and W. V. Quine.

The most influential publication of the core group was a slim pamphlet issued in October 1929 as a kind of manifesto of the movement, titled *The Scientific Conception of the World*.[6] The very title was a trumpet blast in the fight to change the balance again, to put science back at the center of modern culture, and against what the booklet called, in the first sentence, the chief alternative, the tendency toward metaphysical and theologizing thought, those old helpmates of the Romantic movement.

Although most of the scholars involved in the Vienna Circle concerned themselves chiefly with the study of the epistemological and logical problems at the foundations of science, there was a clear undercurrent of wider cultural, social, political, and pedagogic ambitions as well. For, as the manifesto said, "The attention toward questions of life is more closely related to the scientific world conception than it might at first glance appear. . . . For instance, endeavors toward the unification of mankind, toward a reform of school and education, all show an inner link with the scientific world conception. . . . We have to fashion intellectual tools for everyday life. . . . The vitality that shows itself in the efforts for a rational transformation of the social and economic order permeates the movement for a scientific world conception, too" (pp. 304–305).

The members of the circle associated themselves explicitly not with the Platonists and Pythagoreans, but with the Sophists and

Epicureans, "with those who stand for earthly being, and the here and now." A science free from metaphysics would be a unified science; it would know no unsolvable riddles; it would train thinking to produce clear demarcations between meaningless and meaningful discourse, between intellect and emotion, between the areas of scientific scholarship on the one hand and myth on the other. Just as this approach would, by this formulation, clarify the foundations of mathematics, of the physical sciences, of biology and psychology, it would also demystify the foundations of the social sciences, "and in the first place . . . history and economics." The empiricist, antimetaphysical attitude would hasten the rejection of such dangerous conceptions as "folk spirit," and would "liberate one from inhibiting prejudices."

Thus, the "debris of millennia" would be removed, and "a unified picture of this world" would emerge, free from magical beliefs. The social and economic struggles of the time would be ameliorated because the "broad masses of people" would reject the doctrines capable of misleading them (pp. 315–317). Beyond that, the spirit of the scientific world conception would penetrate "in growing measure the forms of personal and public life, in education, upbringing, architecture, and the shaping of economic and social life according to rational principles." And the manifesto for a new modernity ended with the blazing formulation, in italics: *The scientific world conception serves life, and life receives it* " (p. 318).

Perhaps the most carefully developed of the many publications expressing the circle's position on science and its rationality as the keys to a sane world picture was the major book by Richard von Mises, the Austrian scientist, mathematician, engineer, and philosopher (as well as scholar of the poet Rainer Maria Rilke). Von Mises entitled his weighty volume, with a bit of irony, *Kleines Lehrbuch des Positivismus*: The aim was not only to show what an empiricist-rational scientific world conception would consist of, what its tools would be, and what problems it could solve within the sciences, from mathematics and physics to biology and the social sciences.[7] All this is done in great detail; but an equally motivating force was to present thereby a choice from the then-reigning alternatives in German-speaking Europe: the Kantianism in Germany and the clerical-metaphysical trend in Austria, both of which were then being interspersed with the growing totalitarian ideologies. Von Mises noted his quite explicit opposition to what he called "negativism," in which he includes systematic philosophical and political anti-intellectualisms that have remained part of the present landscape. Among the examples he cited were, in fact, Oswald Spengler, and the once-popular German philosopher Ludwig Klages, whose point of view was enshrined even in the title of his main work, *The Mind as Enemy of the Soul.*

As a sign that von Mises's main aim of the book was to put science at the center of a healthy culture in the largest meaning of the term, his volume dealt at length with the way the scientific world conception would illuminate the understanding of metaphysics, poetry, art, the law, and ethics. The underlying commonality of the various forms of cultural achievements was considered by von Mises to be due to the principal unity of their methods if carried through rationally and soundly. The original readers of the book must have felt themselves to be in the presence of an updated follower of Auguste Comte. The very last sentence is, as it were, the summary of the entire project: "We expect from the future that to an ever-increasing extent scientific knowledge, i.e., knowledge formulated in a connectable manner, will regulate life and the conduct of man" (p. 370).[8]

FREUD: INSTINCTUAL PASSIONS VERSUS REASONABLE INTERESTS

But now we shall see the lever of sentiments shift the balance once more, and indeed on the very issue of whether knowledge formulated in a scientific manner can lead mankind to saner and more rational conduct. In 1929, the same year in which the optimistic manifesto of the Vienna Circle was published, Sigmund Freud, writing in the same city, produced a book of his mature years giving his somber and pessimistic answer. To the founder of psychoanalysis, the role of science in human culture had been a continuing preoccupation, and in 1911 he had still been optimistic enough to sign the *Aufruf* of the Society for Positivistic Philosophy. But in that book of late 1929, *Das Unbehagen in der Kultur*,[9] Freud found that science, while counting among the most visible manifestations of civilization, was at best only an ameliorating influence in a titanic struggle on which the fate of the culture depended. That struggle, he said, was centered on mankind's often doomed effort to master "the human instinct of aggression and self-destruction." Even at that time he saw, as expressed in the last paragraph of the book, that "mankind has gained control over the forces of nature to such an extent that with their help it may have no difficulty to exterminate one another to the last man" (p. 92).

Freud held that the restrictions which civilization imposes upon our instinctual urges produce an irremediable antagonism against those fetters. Our innate "Destructive Instinct," or "Death Instinct" (pp. 7, 8), is a drive constantly at odds with the civilizing project to elevate the moral condition of mankind. Freud wrote, ". . . man's natural aggressive instinct, the hostility of each against all, and of all against each, opposes this program of civilization. This aggressive instinct is the derivative and the main representative of the death instinct which we have found alongside of Eros and which shares world-domination with it. And now, I think, the meaning of the evolution of civilization is no longer obscure to us. It must present the struggle between Eros and Death, between the instinct of life (*Lebenstrieb*) and the instinct of destruction (*Destruktionstrieb*), as it works itself out in the human species. This struggle is what all life essentially consists of, and the evolution of civilization may therefore be simply described as the struggle for life of the human species. And it is this battle of the giants that our nursemaids try to appease with their lullaby about Heaven" (p. 69).

In this conflict, scientific and other cultural activities arise as the result of a "sublimation of instinctual aims," making science at first glance merely a "vicissitude which has been forced upon the instincts by civilization." The accomplishments of science and technology originated as welcome tools in the effort to protect men against the hostile forces of nature; they have now become "cultural acquisitions" that "do not only sound like a fairy tale, they are actual fulfillments of every—or almost every—fairy tale wish." They verge on our attaining the old ideas of "omnipotence and omniscience." Man "has, as it were, become a kind of prosthetic God" (pp. 38–39).

But there's the rub: happiness still eludes him. "Present-day man does not feel happy in his God-like character," either individually or in terms of the group. That again has its reason in the fact that "civilization is built upon a renunciation of instinct," such as sexuality and aggressiveness, and "presupposes precisely the nonsatisfaction (by suppression, repression, or some other means) of powerful instincts." Hence the "cultural frustration" (*Unbehagen*) which dominates the entire field of social relationships between human beings (pp. 43–44, 62).

Freud's pessimistic conclusion follows: "In consequence of this primary mutual hostility of human beings, civilized society is perpetually threatened with disintegration. The interest of work in common would not hold it together; instinctual passions are stronger than reasonable interests. . . . In spite of every effort these endeavors of civilization have not so far achieved very much. . . . It is always possible to bind together a considerable number of people in amity, so long as there are other people left over to receive the manifestations of their aggressiveness," as in religious or ethnic persecution (pp. 59, 61).

During the decades since this was written, modern history has all too often seemed to be the experimental verification of Freud's dark assessments, according to which science and all other cultural activities cannot fully displace our animal nature from its reigning position, but can only delay the ultimate fate that threatens our society.

SCIENTISTS AS "BETRAYERS OF THE TRUTH"

Let us now turn to the most recent period. We are familiar enough with the fluctuations, during the 1960s and 1970s, of opinion in academe and among the public regarding the interactions of science and society. But starting in the early 1980s, a new and powerful element entered into this discussion which has now been assuming ever greater attention and institutionalization, at least in the United States. The new element, the new force adding to the derogation of the credibility of science, is the insistence from some quarters—which increasingly has fallen on receptive ears among the population—that to a previously completely unrealized degree the pursuit of science is, and has been all along, ever since the days of Hipparchus and Ptolemy, thoroughly corrupt. Consequently, severe external measures must be applied to the practice of science. This assertion, which has become louder and louder over the past few years in books, official reports, and hundreds of articles, has spawned dramatic public hearings, the formation of special government agencies, university bureaucracies, and quite a few careers. The safeguarding of ethical practices and uses of science, of which there has been a long tradition within the scientific community, is now to be put into better and wiser hands.

A striking, pacesetting example of this assertion was the 1982 book by two influential *New York Times* science editors, William Broad and Nicholas Wade. It states its intention in its title, *Betrayers of the Truth: Fraud and Deceit in the Halls of Science*,[10] and follows up with the unqualified cannon shot of the opening sentence: "This is a book about how science really works." Going far beyond the need to expose the relatively few rotten apples in any barrel, which the scientific community itself has long recognized as an obligation if only for its own health, this kind of rhetoric has become commonplace. As this book and its many followers proclaim, the relatively few, sad cases of real or alleged misbehavior are the litmus test for the entire enterprise; fraud and deceit are depicted as being part of the very structure of scientific research.

Similarly, the report to Congress by the Congressional Research Service, entitled "Scientific Misconduct in Academia," stated that, more and more, "the absence of empirical evidence which clearly indicates that misconduct in science is not a problem . . . suggests that significant misconduct remains a possibility." Among all the targets to preoccupy those who are charged with timely attention to misconduct damaging our republic, this formulation singles out the conduct of science as being guilty until proven innocent. Moreover, the tendency has recently been to include in the allegation of *scientific* misconduct not only falsification of data, plagiarism, and the like, but also the spectrum of misdeeds more common to flawed mankind generally, and for which sanctions have existed, e.g., "use of university resources for inappropriate purposes, sexual harassment, racial discrimination," etc.[11]

Similarly, the Office of Scientific Integrity Review (OSIR) of the Department of Health and Human Services made part of its proposed definition of "misconduct" in science, apart from fabrication, falsification, and plagiarism, "practices that seriously deviate from those that are commonly accepted in the scientific community" (Federal Code: 42 C.F.R. 50.102). The intention here may have been to parallel the way the Supreme Court defined obscenity by reference to the current standards of the local community. However, when it comes to making progress in science, some practices contrary to those common at the time have again and again been the very hallmark of needed innovations—from putting mathematics into physics in the seventeenth century to the introduction of quanta, which pained even the originator, Max Planck himself, and to the more recent innovation of modern teamwork. The proposed definition of misconduct, with its potential for mischief, was one more example of the gap between the culture of science and the culture outside the lab. One should add that to her credit the director of the National Institutes of Health at the time intervened on that point, objecting that such a community standard "would have implicated even the discoverer of penicillin, who serendipitously found good use for bacteria growing in a contaminated lab dish."[12]

The power of the current generalized allegations against the conduct of scientists has two components. The first is of course the astonishing claim that basic research scientists in considerable numbers are intentionally false to their own most fundamental avowed mission, namely, to the pursuit of truths; in other words, that not just a few apples are rotten, but that the entire barrel is.

Without the vastly overblown allegation of pervasive and ingrained fraud and deceit in science, even the presence of the occasional scandalous misdeeds by a relatively small number of the world's millions of scientific researchers would not have been taken so seriously that in the United States the newspapers, college courses, training courses for scientists and physicians, commissions, congressional committees, scientific societies, and so on, have become massively and expensively preoccupied with the institutionalization of prevention of misconduct in science. The unrelenting accounts of specific incidents, some outrageous, more of them sensationalized, have left the public feeling that a great plague of dishonesty has invaded all academic laboratories. As the journal *Nature* noted shrewdly, the current trend is resulting in "a slow—and Hollywood-assisted—erosion of [the scientists'] public image, . . . [replacing it] in the public mind by a money-grabbing plagiarizing con-artist."[13] *Time* magazine chimed in with an essay on scientists today, beginning with, "Scientists, it seems, are becoming the new villains of Western society." A raft of bestselling books add up the allegations in diatribes that have the frank view, in the words of Bryan Appleyard's polemic *Understanding the Present: Science and the Soul of Man*, that science must be "humbled." We are, it appears, standing only on the shoulders of dwarfs. As the distinguished chemist and molecular biologist M. F. Perutz put it in a masterful exposé of the faults in a typical book that recast a pioneer scientist as a perpetrator of fraud and careerism: "The entire approach emphasizing 'relative' truth seems to me a piece of humbug masquerading as an academic discipline; it pretends that its practitioners can set themselves up as judges over scientists whose science they fail to understand."[14]

What is getting lost in this avalanche of excitement, and also in the generally poor, even self-flagellating, responses from most scientific institutions, is a thorough inquiry into the actual rate of serious misconduct among scientists, the kind of empirical research that would yield a reasonable estimate of the relative rate of unacceptable incidents. I have found only some scattered, preliminary steps in this direction, but these suggest that in fact the actual rate of misconduct (rather than suspected, alleged, or "perceived" without hard evidence) is remarkably *low*. Among the available, reasonably quantifiable measures is, for example, the National Library of Medicine finding that for the period of 1977 to 1986, when about 2,780,000 articles were published in the world's biomedical literature, 41 of these had to be withdrawn because fraudulent or falsified data appeared in them—a rate of under two one-thousandths of one percent of scientific publications per decade. Other data support the same point. Thus the Food and Drug Administration, responding to allegations or evidence of misconduct in clinical research with investigational new drugs, submitted twenty cases of suspected fraud or other criminal violations to the U.S. Attorney General's office. These resulted in thirteen convictions of clinical investigators—about one per year, on the average.[15]

Nobody does or should condone even a single case of misconduct. But even if the actual rate were as much as a hundred times greater than these figures indicate, the intellectually most interesting questions would be, first, why science as a whole progresses so well despite being the work of mere human beings; second, how rare the cases of alleged misconduct are in this field compared with those in others, ranging from the world of finance, law, industry, journalism, and government at every level; and third, why the few cases of highly publicized charges of misconduct in science can so severely undermine the trust and confidence of the public and its representatives in the integrity of research in general.

SCIENCE AS MYTH

The answer to those questions is in good part that there is indeed another, reinforcing reason for the widespread success of the assault on the credibility of scientific research. This second line of attack has been opened up by a loose assemblage made up of a branch of contemporary philosophy of science, the so-called "strong-program" constructivist portion of sociology, a subset of the media, a small but growing number of governmental officials and political aspirants, and a vocal segment of literary critics and political commentators associated with the avant-garde of the postmodern movement. This is a potent and eloquent collective of just the sort that in the past has successfully challenged the prevailing worldview.

The overall message evolving over the past decade or two from that direction is no longer based only on stories of unacceptable behavior among a few scientists. The charge has been generalized and made even more serious: Put in starkest terms, the claim is that the most basic fraud committed by the members of the scientific community is their assertion *that there are any truths to be found at all*. For there really is nothing there even to betray and falsify; and consequently, science is inherently not corrigible, even if all misconduct were eliminated.

From that point of view, the business of science is mainly careerist: for example, building and operating expensive institutions that claim to be looking for objectively ascertainable information about entities such as quarks and bosons—which, however, are nothing more than "socially constructed" fictions. Against the naïve realism that most scientists still embrace, and the agnosticism of the more sophisticated ones, the new critics counterpoise the radical solution: as one sociologist of science put it recently, "There is no Nature; there is only a communication network [among scientists]." The literature in academe is now full of statements such as "science is a useful myth," or "we must abolish the distinction between science and fiction," "science is politics by other means," or "the search for knowledge is driven by the desire for power."[16]

Scientists have tended to adopt the Baconian view that the acquisition of basic knowledge of causes and interrelations of phenomena—by processes not easily predictable or fully understood—can yield power over those of nature's forces which cause our burdens and ills. But now, the new consortium tells us, the arrow really goes the other way: not from knowledge to power, but from power to knowledge, and to a rather questionable knowledge at that. The attempts to find generally applicable, shareable knowledge about what might be called reality—through the use of both the rational and the intuitive faculties of individual scientists, through the positing of freely chosen concepts that later can be tested for merit (as Einstein will show us in Part Two of this book), and through the skeptical but collaborative attempt to achieve consensus—are not only disguised as doomed exercises, but are said to have led ironically to the disasters that have marked the twentieth century. The modern era, launched under the flag of progress, has only led to tragedy. The extreme overoptimism of a Herbert Spencer or a Friedrich Engels can never be replaced by a soberer conception. Progress is illusion. The globalizing program of science—to find basic unities and harmony transcending the level of apparent variety and discord—is held to be completely contrary to the postmodern drive, which celebrates individual variety and the equality of standing of every conceivable style and utterance, every group and competing interest. Ours is the time to face the end of the search for foundations, the "End of the Modern Era"; we are in a state called the "objectivity crisis"—a fashionable phrase found in the titles of learned conferences as well as in policy-setting documents to be examined shortly.

Together, these slogans of the newly emerging sentiment indicate that the aim is not merely a call for the improvement of practice or for increased accountability, which is appropriate and being pursued through earnest actions, but is at bottom, for the main branch of the movement of critics, the delegitimation of science as one of the valid intellectual forces, a reshaping of the cultural balance, as we shall see in more detail below. Even the most irrational fringe borrows from the ideas, widespread in parts of academe, that technology is inherently a source of "disaster for the human race," and that the scientist's "motive is neither curiosity nor a desire to benefit humanity but the need to go through the power process," to quote from the lengthy manifesto "Industrial Society and Its Future," written by the so-called Unabomber. (When the FBI circulated this manifesto among academics for their reactions, the *Chronicle of Higher Education* reported, "Some historians of science said that except for his endorsement of violence, the bomber's arguments are consistent with the critical strain of scholarly thought in the field," and quoted one scholar who observed that except for the call for violence, "it's a sensible academic piece.")

There is a big difference in the current attack against science and the history of *internal* movements of protest, such as those of the logical positivists within philosophy, the impressionists or dadaists within art, the modern composers within music, etc. In all those cases, it was some of the best talent in the field that took up the task of renewal. Not so here—the motivating force is not renewal from within, but radical cultural politics from without.[17]

THE ROMANTIC MOVEMENT'S CHALLENGE

Here we meet a clarifying fact: The contest before us is not new, but draws on historic forces of great strength and durability. Therefore it will be instructive to trace some of the individual steps and stages in this remarkable development, so as to make it easier to extrapolate and to preview the new terrain we may have before us. While I can here only point briefly to a few recent milestones, I shall seek documentation in the recent writings of some of the most distinguished thinkers, rather than, say, through representatives of the Dionysian undercurrent.

Our first informant and guide is Isaiah Berlin, widely regarded as a most sensitive and humane historian of ideas. The collection of his essays, published as the fifth volume of his collected papers,[18] opens with a startling dichotomy. He writes: "There are, in my view, two factors that, above all others, have shaped human history in this century. One is the development of the natural sciences and technology, certainly the greatest success story of our time—to this great and mounting attention has been paid from all quarters. The other, without doubt, consists of the great ideological storms that have altered the lives of virtually all mankind: the Russian revolution and its aftermath—totalitarian tyrannies of both right and left and the explosion of nationalism, racism and, in places, of religious bigotry, which interestingly enough, not one among the most perceptive social thinkers of the nineteenth century had ever predicted" (p. 1). He adds that if mankind survives, in two or three centuries' time these two phenomena will "be held to be the outstanding characteristics of our century, the most demanding of explanation and analysis."

What might the author intend by so juxtaposing these two "great movements"? One's first temptation may be to see a connection through the fact that during World War II the ingenuity and frantic work of scientists among the Allies, supporting the valor of the Allied soldiers, brought an end to the totalitarian tyranny of that period, which might well have triumphed over the democracies and established itself at least throughout Europe.

But such a response would not be to the point here. What is on Isaiah Berlin's mind is quite different. As we follow his eloquent and subtle analysis, it dawns on the reader that science and tyranny, the two polar opposite movements which he holds to have defined and shaped the history of the twentieth century, are somehow intertwined—that the development of the modern natural sciences and technology may, *through the reactions against them*, have unintentionally and indirectly contributed to the rise of "totalitarian tyrannies."

This stunning connection, to be sure, is never explicitly spelled out by the author. But we can glimpse the implicit argument later in the book, in his chapter significantly entitled "The Apotheosis of the Romantic Will: The Revolt against the Myth of an Ideal World." There, Berlin summarizes the chronology of some basic concepts and categories in the Western world, specifically the changes in "secular values, ideals, [and] goals." What commands his attention is the change away from the belief in the "central core of the intellectual tradition . . . since Plato," and toward a "deep and radical revolt against the central tradition of Western thought" (p. 208), a revolt which in recent times has been trying to wrench Western consciousness into a new path.

The central core of the old belief system, one that lasted into the twentieth century, rested on three dogmas that the author summarized roughly as follows. The first is that "to all genuine questions there is one true answer, all others being false, and this applies equally to questions of conduct and feeling, to questions of theory and observation, to questions of value no less than to those of fact." The second dogma is that "the true answers to such questions are in principle knowable." And the third: "These true answers cannot clash with one another." They cannot be incommensurate, but "must form a harmonious whole," the wholeness being assured by either the internal logic among the elements, or their complete compatibility (pp. 209–211).

Out of these three ancient dogmas both institutionalized religions and the sciences developed to their present form (although one might add that modern scientists, in their practice, have become aware of the need of proceeding antidogmatically, by conjecture, test, refutation, and assaying probability). In their pure state, these systems are utopian in principle, for they are imbued by the optimistic belief, inherent in and derivable from the dogmas, that "a life formed according to the true answers would constitute the ideal society, the golden age." All utopias, Isaiah Berlin reminds us, are "based upon the discoverability and harmony of objectively true ends, true for all men, at all times and places"—and by implication the same is true for scientific and technical progress, which are aspects of our drive toward what he calls "a total solution: that in the fullness of time, whether by the will of God or by human effort, the reign of irrationality, injustice and misery will end; man will be liberated, and will no longer be the plaything of forces beyond his control [such as] savage nature. . . ." This is the common ground shared by Epicurus and Marx, Bacon and Condorcet, *The Communist Manifesto*, the modern technocrats, and the "seekers after alternative societies" (pp. 212–213).

But, Isaiah Berlin now explains, this prominent component of the modern world picture is precisely what was rejected in a revolt by a two-centuries-old countermovement that has been termed Romanticism or the Romantic Rebellion. From its start in the German *Sturm und Drang* movement of the end of the eighteenth century, it grew rapidly in Western civilization, vowing to replace the ideals of the optimistic program based on rationality and objectively true ends, by the "enthronement of the will of individuals or classes, [with] the rejection of reason and order as being prison houses of the spirit."

My own favorite summary of the negative view of science in nineteenth-century literature is the antihero in Ivan Turgenev's gripping novel, *Fathers and Sons*. One of the greatest figures of Russian literature, together with Gogol, Dostoyevsky, and Tolstoy, Turgenev was a poet largely in the tradition of nineteenth-century Romanticism, inspired by Goethe, Schiller, and Byron, among others. *Fathers and Sons* was published in 1861. Its main figure is Yevgeny Vassilevich Bazarov, a university student of the natural sciences, expecting shortly to get his degree as a physician. Being a scientist who "examines everything from a critical point of view," he confesses himself also to be ideologically and politically a nihilist, the natural consequence of not acknowledging any external authority. All talk of love, or the "mystic relationship between a man and a woman," is to him just "romanticism, humbug, rot, art." It would be better to study the behavior of beetles. Even on his vacation he has brought along a microscope and fusses over it "for hours at a time." Reading Pushkin, he says, is for little boys. He thinks it would be much better to start with Ludwig Büchner's *Force and Matter*, a book published in 1855 and embodying such a flagrantly materialistic view that Büchner was forced to resign from his professorship in Germany. (It is, as it turned out later, the very book Albert Einstein singled out in his *Autobiographical Notes* as one of the two or three that most impressed him as a boy, and caused him to turn to the pursuit of science.)

What matters, Bazarov claims, "is that two and two are four—all the rest is nonsense." When he meets a clever and beautiful woman, he startles his friend by saying that hers would be a beautiful body to examine—on a dissection table. As if in revenge, fate brings him to the bedside of a villager dying of typhus, and he is made to help in the postmortem. But he cuts himself with his scalpel, and soon he is on the verge of delirium, a victim of surgical poisoning. As he is dying, he tries to keep hold on reality by asking himself aloud, "Now, what is 8 minus 10?" In short, he is a caricature recognizable throughout literature, devoid of the spectrum of imaginative tools one actually needs to do science, as we shall see later on in this book—except that typically in literature the emotionally dysfunctional scientist, from Dr. Frankenstein to Dr. Strangelove, causes surgical sepsis not only in himself but in all those around him.

Returning to Isaiah Berlin's account, it is striking that, as he notes, no one predicted that the current form of the worldwide Romantic Rebellion would be what dominates "the last third of the twentieth century." The Enlightenment's search for generalizability and rational order is depicted by the rebels of our time as leading at best to the pathetic Bazarovs of science, and those must be replaced by the celebration of the individual, by flamboyant antirationalism, by "resistance to external force, social or natural." In the words of Johann Gottfried von Herder, the rebel shouts: "I am not here to think but to be, feel, live!" (p. 223). Truth, authority, and nobility come from having heroically suffered victimization.

This assertion of the individual will over shareable reason has undermined what Isaiah Berlin called the pillars of the Western tradition. The Romantic Rebellion of course has also given us enduring masterpieces of art, music, and literature. But it originated, as it were, as an antithetical mirror image, created in reaction to the very existence of the earlier Enlightenment-based conception. In the apotheosis of the Romantic Will in our time, it glows forth as the alternative, the "romantic self-assertion, nationalism, the worship of heroes and leaders, and in the end . . . Fascism and brutal irrationalism and the oppression of minorities" (p. 225). Moreover, in the absence of "objective rules," the new rules are those that the rebels themselves make: "Ends are not . . . objective values. . . . Ends are not discovered at all but made, not found but created."

As a result, Berlin writes, "this war upon the objective world, upon the very notion of objectivity," launched by philosophers and also through novels and plays, infected the modern worldview; the "romantics have dealt a fatal blow" to the earlier certainties, and have "permanently shaken the faith in universal, objective truth in matters of conduct" (pp. 236–237)—and, he might have added, in science as well. As any revolt does, this one puts before us seemingly mutually incompatible choices. Just as with quite opposite cases of excess such as Ostwald's, it is again Either/Or. Lost from sight in this combat is the needed complementarity of mankind's rational, passionate, intuitive, and spiritual functions—a complementarity on which, as we shall see exemplified in Part Two of this book, good work in science itself depends. But one is reminded here of the fact that violent extremes tend to meet. Thus the poet William Blake, the epitome of the Romantic Rebellion—who called Bacon's, Newton's, and Locke's work satanic—composed in his *The Marriage of Heaven and Hell* (1790) one of the "Proverbs" that reveal the outrageous credo of so many of the opposing actors in this story to this day: "*The road of excess leads to the palace of wisdom.*"

THE ROMANTIC REBELLION INFUSES STATE POLICY

Other authors provide verification and elaboration of the implications of Berlin's findings, and especially so of what in my view is the foremost danger posed by the movement: *the ominous joining* in *the twentieth century of the extremes of a Romantic Rebellion with irrational political doctrines.* This was evident in the Cultural Revolution in Mao's China, in the USSR, and in other totalitarian systems. To glance at one telling example, the historian Fritz Stern has written about the early phases of the growth of Nazism in Germany, when there arose in the 1920s, in his words, the "cultural Luddites, who in their resentment of modernity sought to smash the whole machinery of culture." The fury over an essential part of the program of modernity, "the growing power of liberalism and secularism," directed itself naturally against science itself. Julius Langbehn was one of the most widely read German ideologues in the 1920s, and Stern writes of him, "Hatred of science dominated all of Langbehn's thought. . . . To Langbehn, science signified positivism, rationalism, empiricism, mechanistic materialism, technology, skepticism, dogmatism, and specialization. . . ."

Long before the Nazis assumed governmental power, some German scientists and other scholars demanded that a new science be created to take the place of the old one which they discredited—a new "Aryan science," based on intuitive concepts rather than those derived from theory; on the ether, the presumed residence of the spirit, the "*Geist*"; on the refusal to accept formalistic or abstract conceptions, which they reviled as earmarks of "Jewish science"; and on the adoption as far as possible of basic advances "made by Germans."

In a classic study, Alan Beyerchen identified some of the other main pillars of Aryan science.[19] There we find themes uncomfortably similar to those that are again fashionable. A prominent part of Aryan science was, of course, that science is, as some would now say, basically a social construct, so that the racial heritage of the observer "directly affected the perspective of his work." Scientists of undesirable races, therefore, could not qualify; rather, one had to listen only to those who were in harmony with the masses, the "*Volk*." Moreover, this *völkisch* outlook encouraged the use of ideologically screened nonexperts to participate in judgments on technical matters (as in the *Volksgerichte*). The international character of the consensus mechanism for finding agreement in science was also abhorrent to the Nazi ideologues. Mechanistic materialism, denounced as the foundation of Marxism, was to be purged from science, and physics was to be reinterpreted to be connected not with the matter but with the spirit. "The Aryan physics adherents thus ruled out objectivity and internationality in science. . . .

Objectivity in science was merely a slogan invented by professors to protect their interests." Hermann Rauschning, president of the Danzig Senate, quoted Adolf Hitler as follows:

> We stand at the end of the Age of Reason. . . . A new era of the magical explanation of the world is rising, an explanation based on will rather than knowledge. There is no truth, in either the moral or the scientific sense. . . . Science is a social phenomenon, and like all those, is limited by the usefulness or harm it causes. With the slogan of objective science the professoriat only wanted to free itself from the very necessary supervision by the State.
>
> That which is called the crisis of science is nothing more than the gentlemen are beginning to see on their own how they have gotten onto the wrong track with their objectivity and autonomy.[20]

One issue was how technology, so useful to the state, could be fitted into the Romantic idea. In recent times, many antimodern movements, including fundamentalist ones, have embraced technology. The physicist Philipp Lenard, a chief cultural hero of Nazi propaganda, spoke for at least a minority when he said that the tendency of scientific results to prepare the ground for practical advances has led to a dangerous notion, that of man's "mastery" of nature: Such an attitude, he held, only revealed the influence of "spiritually impoverished grand technicians" and their "all-undermining alien spirit." This idea, too, had its roots in the centuries-old history of the rise of Romantic thought. Alan Beyerchen provides us with a summary with his observation that "the romantic rejection of mechanistic materialism, rationalism, theory and abstraction, objectivity, and specialization had long been linked with beliefs in an organic universe, with stress on mystery [and] subjectivity. . . ."

All these excesses were couched in phrases reminiscent of those currently used to delegitimate the intellectual authority of science. But it is necessary to keep in mind that the common ancestry of these views does not mean that there is necessarily a causal connection between them. Today's antiscientists usually do not know how closely they may be following the historical precedents. This applies also to the next case, as I turn now to the position embraced by another distinguished contemporary icon among humanists, although an advocate rather than an analyst. His writings on this topic are—like those of Oswald Spengler, or the positivists—of interest here not because they now represent majority positions, which they do not, but because they have the potential for wide resonance at a turning point of sentiments. Also, in this case we shall see that the relation between modern natural science and the rise of totalitarianism, which Isaiah Berlin considered to be only the result of an obscene historic counterreaction, receives now a much more sinister interpretation: the two become directly, causally linked.

This ominous linkage has been argued repeatedly in writings over the past ten years by Václav Havel, the Czech poet, playwright, resistance fighter against Marxist-Leninist oppression, and statesman. In the passages to be discussed, we will notice that Havel subscribes to many of the themes discussed in Isaiah Berlin's analysis; but Havel's key point is that totalitarianism in our time was the perverse end result of a trend of ideas embodied in the program of science itself. In this sense, Western science gave birth to communism; and with the fall of the latter, the former has also been irremediably compromised.

Looking back on the twentieth century, other Central Europeans might characterize the tragedies of our age as due to the forces of brutal irrationality and bestiality, a reversion to ruthless autocracies in which the fates of millions were sealed by the whims of Kaiser Wilhelm, Hitler, Stalin, and their henchmen—rather than being the result of organized skepticism and the search for reasoned consensus, which are at the heart of science. But Havel finds the chief sources of trouble in the twentieth century to have been the very opposite, namely, the habit—in his words—of "rational, cognitive thinking," "depersonalized objectivity," and "the cult of objectivity." He advises us to take refuge now in unrepeatable personal experience, in intuition and mystery and the other mainstays of the Romantic Rebellion. I must let him put his case at some length in his own words; for while he eschews the documentation or balanced account of the scholar, he is instead in fine command of the rhetoric of persuasion, the ease of unspecified assertions and generalizations, and of the chief art of the dramatist, the suspension of disbelief. The result, for many of his readers, has been hypnotic acquiescence that does not seek to question the generalities and leaps in the prose. The "end of Communism," Havel writes in one of his most widely quoted essays,

> has brought an end not just to the 19th and 20th centuries, but to the modern age as a whole.
>
> The modern era has been dominated by the culminating belief, expressed in different forms, that the world—and Being as such—is a wholly knowable system governed by a finite number of universal laws that man can grasp and rationally direct for his own benefit. This era, beginning in the Renaissance and developing from the Enlightenment to socialism, from positivism to scientism, from the Industrial Revolution to the information revolution, was characterized by rapid advances in rational, cognitive thinking. This, in turn, gave rise to the proud belief that man, as the pinnacle of everything that exists, was capable of objectively describing, explaining and controlling everything that exists, and of possessing the one and only truth about the world. It was an era in which there was a cult of depersonalized objectivity, an era in which objective knowledge was amassed and technologically exploited, an era of systems, institutions, mechanisms and statistical averages. It was an era of freely transferable, existentially ungrounded information. It was an era of ideologies, doctrines, interpretations of reality, an era in which the goal was to find a universal theory of the world, and thus a universal key to unlock its prosperity.
>
> Communism was the perverse extreme of this trend. . . . The fall of Communism can be regarded as a sign that modern thought—based on the premise that the world is objectively knowable, and that the knowledge so obtained can be absolutely generalized—has come to a final crisis. This era has created the first global, or planetary, technical civilization, but it has reached the limit of its potential, the point beyond which the abyss begins.
>
> Traditional science, with its usual coolness, can describe the different ways we might destroy ourselves, but it cannot offer truly effective and practicable instructions on how to avert them.[21]

A listener might object that these passages are built on immense overgeneralizations and illogical jumps, just as flawed as those of the extreme Monists were on the other side; or that at least on factual grounds the self-designation of communist ideology as "scientific" was indeed a fraud. On this last point, the scholar of the history and philosophy of Soviet science, Loren Graham, made the trenchant observation: "In 1992, the playwright and President of independent Czechoslovakia Václav Havel wrote that the fall of communism marked the end of an era, the demise of thought based on scientific objectivity. . . . Was the building of the White Sea Canal in the wrong place and by the most primitive methods, at the cost of hundreds of thousands of prisoners' lives, the blossoming of rationality? Was the disregard of the best technical specialists' advice in the construction of Magnitogorsk, the Dnieper dam and the Baikal-Amur Railway a similar victory for objectivity? Was the education of the largest army of engineers the world has ever seen—people who would come to rule the entire Soviet bureaucracy—in such a way that they knew almost nothing of modern economics and politics an achievement of science? . . . . And even long after the death of Stalin, into the 1980s, what was the Soviet insistence on maintaining inefficient state farms and giant state factories, if not an expression of willful dogmatism that flew in the face of a mountain of empirical data?"22

Yet one may doubt whether Havel would reconsider his position, for the object of his essay is his conclusion, presenting the "way out of the crisis of objectivism," as Havel labels it. Only a radical change in man's attitude toward the world will serve. Instead of the generalizing and objectifying methods that yield shareable, repeatable, inter- or trans-subjective explanations, we must now turn, he says, to the very opposite, which presumably "science" somehow has totally banished from this world, i.e., to "such forces as a natural, unique and unrepeatable experience of the world, an elementary sense of justice, the ability to see things as others do . . . courage, compassion, and faith in the importance of particular measures that do not aspire to be a universal key to salvation. . . . We must see the pluralism of the world. . . . We must try harder to understand than to explain." Man needs "individual spirituality, firsthand personal insight into things . . . and above all trust in his own subjectivity as his principal link with the subjectivity of the world. . . ."

Despite Havel's hint, in passing, of a possible blending of the "construction of universal systemic solutions" or "scientific representation and analysis" with the authority of "personal experience," so as to achieve a "new, postmodern face" for politics, Havel's identification of the "End of the Modern Era" is not to be understood as a reasonable plea for some compromise or coexistence among the rival constructs; that much was announced in an earlier and even sharper version of his essay, one which dealt with the place of modern science quite unambiguously and hence deserves careful reading:

> [Ours is] an epoch which denies the binding importance of personal experience—including the experience of mystery and of the absolute—and displaces the personally experienced absolute as the measure of the world with a new, man-made absolute, devoid of mystery, free of the 'whims' of subjectivity and, as such, impersonal and inhuman. It is the absolute of so-called objectivity: the objective, rational cognition of the scientific model of the world.
>
> Modern science, constructing its universally valid image of the world, thus crashes through the bounds of the natural world which it can understand only as a prison of prejudices from which we must break out into the light of objectively verified truth. . . . With that, of course, it abolishes as mere fiction even the innermost foundation of our natural world. It kills God and takes his place on the vacant throne, so that henceforth it would be science which would hold the order of being in its hand as its sole legitimate guardian and be the sole legitimate arbiter of all relevant truth. For after all, it is only science that rises above all individual subjective truths and replaces them with a superior, trans-subjective, trans-personal truth which is truly objective and universal.
>
> Modern rationalism and modern science, through the work of man that, as all human works, developed within our natural world, now systematically leave it behind, deny it, degrade and defame it— and, of course, at the same time colonize it.[23]

Here we see the giant step which Havel has taken beyond Berlin's analysis: It is modern science itself that has been the fatal agent of the modern era; as if to answer Ostwald's excesses, it is held responsible even for deicide.

Many have been moved by Havel's powerful mixture of poetical feeling, theatrical flourish, and the bold waving of an ancient, blood-stained shirt. The summary of his ideas, published conspicuously under the title "The End of the Modern Era,"[24] made an immediate and uncritical impression on readers of the most varied backgrounds. Among them was one person especially well placed to ponder the values of science, and to draw conclusions of great import for the life of science in the United States. Here we arrive at the last stage on the road to the current understanding of the place of science in our culture.

The person so deeply affected by Havel's piece was none other than a distinguished chairman of the U.S. Congress Committee on Science, Space and Technology, and one of the staunchest and most effective advocates of science during his long tenure in the House of Representatives: George E. Brown, Jr., of California. In 1992 Brown acknowledged that he had received "inspiration" from Havel's essay, "The End of the Modern Era," and decided to reconsider his role as a public advocate of science. He therefore first wrote a long and introspective essay[25] under the title "The Objectivity Crisis," and then presented it to a group of social scientists in a public session at the annual meeting of the American Association for the Advancement of Science, under the title "The Objectivity Crisis: Rethinking the Role of Science in Society."[26]

Persuaded by Havel's version of the Romantic Revolt, Brown cast about earnestly for the consequences it should have for the pursuit of science in the United States. As a pragmatic political leader, he was primarily concerned with how scientific activity may hold on to some legitimacy—by service to the nation in terms of visible "sustainable advances in the quality of life," "the desire to achieve justice" (which he says "is considered outside the realm of scientific considerations"), and all the other "real, subjective problems that face mankind." He now saw little evidence that "objective scientific knowledge leads to subjective benefits for humanity." The privileging of the claim of unfettered basic research is void too, he said, because all research choices are "contextual" and subject to the "momentum of history."

Moreover, science has usurped primacy "over other types of cognition and experience." Here Brown quoted Havel's definition of the "crisis of objectivity" as being the result of the alleged subjugation of our subjective humanity, our "sense of justice, ... archetypal wisdom, good taste, courage, compassion, and faith," to the processes of science, which "not only cannot help us distinguish between good and bad, but strongly assert that its results are, and should be, value free." In sum, Brown held, it would be all too easy to support more research when the proper solution is instead "to change ourselves." Indeed, Brown came to the conclusion that "the promise of science may be at the root of our problems." To be sure, the energies of scientists might still find use if they were properly directed, chiefly into the field of education or into work toward "specific goals that define an overall context for research," such as population control. Embracing a form of Baconianism, Brown thus rejected Vannevar Bush's more general vision for science, a rejection I quoted near the beginning of this chapter (see note 1.) Like Havel's, his answer to the question of whether science can share a place at the center of modern culture was clearly *No.*

When George Brown presented his ideas to an audience of scientists at the session he had organized and for which he had selected a panel of social scientists,[27] only one of the panelists allowed himself to disagree openly; while another of the panelists urged Brown to go even further still: Perhaps not realizing how close he was coming to the "*völkische*" solution tried earlier elsewhere, including in Mao's Cultural Revolution, he seriously suggested that to screen proposals for scientific research funding, the federal government should form a variation of the National Science Foundation's board whose membership should contain such nonexperts as "a homeless person [and] a member of an urban gang." No one there dared to raise an audible objection. One felt as if one glimpsed the shape of a possible future. But it is gratifying to note also that in early 1994, Mr. Brown was apparently moved by the intellectual objections, such as those given above, when they were voiced to him by one or two scientists. He distanced himself from Havel's position.[28] Yet when the new Congress, elected in 1994, drastically reversed the support science had enjoyed for five decades, it was a scenario along Havel's lines rather than Vannevar Bush's. Moreover, one highly placed Congressman of the new majority announced that funding for the gathering of scientific data should be eliminated because such information often leads to Congress adopting regulations, such as those on the protection of the environment or the workplace, which are found burdensome by those whose interests he apparently values more.

In this overview, ranging from the trembling pillars of the Platonic tradition of the West to today's so-called "End of the Modern Era" and the "End of Progress," we have identified some of the chief historic trends that have risen and fallen and risen again in the dynamic out of which the predominant view of an epoch emerges. Today's version of the Romantic Rebellion, while strong in other fields, represents still only a seductive minority view among analysts and science policy makers. It comes not up from the grass roots but down from the tree-tops. However, while it is held among prominent persons who can indeed influence the direction of a cultural shift, the scientists at large, and especially the scientific establishment, have chosen to respond so far mostly with quiet acquiescence. If those trends should continue, and the self-designated postmodernists rise to controlling force, the new sensibility in the era to come will be very different indeed from the recently dominant one.

Experts in science policy are now debating what they call the ongoing renegotiation of the "social contract" between science and society.[29] One can argue that such a change has been overdue for many reasons, one being that the relatively protected position given to science for a good part of the last five decades has less to do with society's commitment than with the Cold War and with the implicit over-promises regarding spin-offs which, as Don K. Price warned long ago,[30] would eventually come back to haunt scientists. Adding concerns about the state of the economy, competitiveness, the lack of general scientific literacy, etc., there is much in such a list to help explain the public's readiness for a reappraisal. But by my analysis, such factors act only as catalysts or facilitators of the tidal change that historically is always potentially present in our culture.

Of course, it may turn out that the present version of the Romantic Rebellion will peter out—although I doubt it will. Or it may gain strength, as it did in the nineteenth century and again at various times in the twentieth, especially when the scientific community paid little attention to the course of events. Or at best a new accommodation might gradually emerge, a "third way," based on a concept analogous to complementarity. That is, it may at last be more widely recognized, by intellectuals and the masses alike, that the scientific and humanistic aspects of our culture do not have to be opposing worldviews that must compete for exclusive dominance, but are in fact complementary aspects of our humanity that can and do coexist productively (as Samuel Taylor Coleridge put it memorably in chapter 14 of his *Biographia Literaria*: "in the balance or reconciliation of opposite or discordant qualities"). At any rate, historians will watch the next stages of the old struggle to define the place of science in our culture with undiminished fascination—although also with an uneasy recollection of Oswald Spengler's prophecy for our *fin de siècle*, of Sigmund Freud's pessimism, and of Isaiah Berlin's analysis of the trajectory of our modern era.

Chapter 2

THE PUBLIC IMAGE OF SCIENCE

WHEN FUTURE GENERATIONS look back to our day, they will envy our generation for having lived at a time of brilliant achievement in many fields, and not least in science and technology. We are at the threshold of basic knowledge concerning the origins of life and the universe itself. We are near an understanding of the fundamental constituents of matter, of the process by which the brain works, and of the factors governing behavior. We launched the physical exploration of space and have begun to see how to conquer hunger and disease on a large scale. Scientific thought appears to be applicable to an ever wider range of studies. With current technical ingenuity one could in principle hope to implement many of the utopian dreams of the past.

Hand in hand with the quality of scientific work today goes an astonishing quantity. The worldwide scientific output is vast. And the amount of work being done is increasing at a rapid rate, doubling approximately every ten or twenty years. Every phase of daily and national life is being penetrated by some aspect of this exponentially growing activity.

It is appropriate, therefore, that searching questions are now being asked about the function and place of this lusty giant. Just as a person's vigorously pink complexion may alert the trained eye to a grave disease of the circulatory system, so too may a local celebration of the success and growth of science and technology turn out, on more thorough study, to mask a deep affliction in our culture. And indeed, anyone committed to the view that science should be a basic part of our intellectual tradition will soon find grounds for concern. Even among educators, scholars, and commentators of our culture, one now hears all too often scientific research described as being an unpleasant, soulless activity, merely "logical," "linear," "hierarchical," and devoid of all human passion. Any practicing scientist knows this to be an absurd characterization, one that at best might be excused as an opinion formed while taking a bad science course—if any—in school.

Some of the major indicators of the relatively narrow place science, as properly understood, occupies in the total culture are quantitative. A nationwide survey found that nearly 40 percent of the men and women who had attended college in the United States confessed that they had not taken a single course in the physical and biological sciences. Similarly, the mass media pay only negligible attention to the subject: the newspapers have been found to give less than 5 percent of their (nonadvertising) space to factual presentations of science, technology, and medicine; and television devotes even less. In short, all our voracious consumption of technological devices, all our talk about the threats or beauties of science, and all our money spent on engineering development should not draw attention away from the fact that the pursuit of scientific knowledge itself is not a strong component of the operative system of general values.

THE ATOMIZATION OF LOYALTIES

In the qualitative sense, and particularly among intellectuals, the indicators are no better. One hears talk of the hope that the forces of science may be tamed and harnessed to the general advance of ideas, that the much deplored gap between scientists and humanists may be bridged. But the truth is that both the hopes and the bridges are illusory. The separation—which I shall examine further—between the work of the scientist on the one hand and that of the intellectual outside science on the other is increasing, and the genuine acceptance of science as a valid part of culture is becoming less rather than more likely.

Moreover, there appears at present no force in our cultural dynamics strong enough to change this trend. This is due mainly to the atrophy of two mechanisms by which the schism was averted in the past. First, the common core of their early education and the wide range of their interests were apt to bring scholars and scientists together at some level where there could be mutual communication on the subjects of their individual competence; and second, the concepts and attitudes of contemporary science were made a part of the general humanistic concerns of the time. In this way a reasonable equilibrium of compatible interpretations was felt to exist, during the last century, between the concepts and problems of science on the one hand and of intelligent common sense on the other; this was also true with respect to the scientific and nonscientific aspects of the training of intellectuals. Specialists, of course, have always complained of being inadequately appreciated; what is more, they are usually right. But although there were some large blind spots and some bitter quarrels, the two sides were not, as they are now in danger of coming to be, separated by a gulf of ignorance and indifference.

It is not my purpose here to urge better science education at the expense of humanistic and social studies. On the contrary, the latter do not fare much better than science does, and the shabby effort devoted to science is merely the symptom of a more extensive sickness of our educational systems. Nor do I want to place all blame on educators and publicists. Too many scientists have forgotten that, especially at a time of rapid expansion of knowledge, they have a special obligation and opportunity with respect to the wider public, and that some of the foremost researchers took great pains to write expositions of the essence of their discoveries in a form intended to be accessible to the nonscientist. In the humanities, too many contributors and interpreters seem to scoff at Shelley's contention in his *Defence of Poetry* that one of the artist's tasks is to "absorb the new knowledge of the sciences and assimilate it to human needs, color it with human passions, transform it into the blood and bone of human nature."

It is through the accumulation of such neglects, just as much as through deterioration in the quantity and quality of instruction, that the place of science as a meaningful component of our culture has been so easily shaken again by the Romantic Rebellion discussed in Chapter 1. Again, this process is to a large extent one aspect of the increasing atomization of loyalties within the intelligentsia. The writer, the scholar, the scientist, the engineer, the teacher, the lawyer, the politician, the physician—each now regards himself or herself first of all as a member of a separate, special group of fellow professionals to which almost all allegiance and energy are given; only very rarely does the professional feel a sense of responsibility toward, or of belonging to, a larger intellectual community. This loss of cohesive links is perhaps the most relevant symptom of the disease of our culture, for it points directly to one of the disease's specific causes: a failure of image.

PURE THOUGHT AND PRACTICAL POWER

Each person's image of science may differ in detail from that of the next, but all public images are in the main based on one or more of seven general positions. The first of these goes back to Plato and portrays science as an activity with double benefits: science as pure thought helps the mind find truth, and science as power provides tools for effective action. In book 7 of the *Republic*, Socrates tells Glaucon why the young rulers in the ideal state should study mathematics: "This, then, is knowledge of the kind we are seeking, having a double use, military and philosophical; for the man of war must learn the art of number, or he will not know how to array his troops; and the philosopher also, because he has to rise out of the sea of change and lay hold of true being. . . . This will be the easiest way for the soul to pass from becoming to truth and being."

The main flaw in this image is that it omits a third vital aspect. Science has always had in addition a metaphoric function—that is, it generates an important part of a culture's symbolic vocabulary and provides some of the metaphysical bases and philosophical orientations of our ideology. As a consequence the methods of argument of science, its conceptions and its models, have permeated first the intellectual life of the time, then the tenets and usages of everyday life. All philosophies share with science the need to work with concepts such as space, time, quantity, matter, order, law, causality, verification, reality. Our language of ideas, for example, owes a great debt to statistics, hydraulics, and the model of the solar system. These have furnished powerful analogies in many fields of study. Guiding ideas—such as conditions of equilibrium, centrifugal and centripetal forces, conservation laws, feedback, invariance, complementarity—enrich the general storehouse of imaginative tools of thought. Ideas emerging from science are, and will continue to be as they have been since the seventeenth century, a central part of modern culture—through pure thought, through practical power, and through metaphoric influence. A sound image of science must embrace each of the three functions; but usually only one of the three is recognized. For example, folklore often depicts the life of the scientist either as one of pure thought—isolated from practical matters and from its metaphoric meaning—or, at another extreme, as dedicated chiefly to technological improvements.

ICONOCLASM

A second image of long standing is that of the scientist as iconoclast. Indeed, almost every major scientific advance has been interpreted— either triumphantly or with apprehension—as a blow against religion. To some extent this position was fostered by the ancient tendency to prove the existence of God by pointing to problems which science could not solve at the time. Newton thought that the regularities and stability of the solar system proved it "could only proceed from the counsel and dominion of an intelligent and powerful Being," and the same attitude governed thought concerning the earth's formation before the theory of geological evolution, concerning the descent of man before the theory of biological evolution, and concerning the origin of our galaxy before modern cosmology. The advance of knowledge therefore made inevitable an apparent conflict between science and religion. It is now clear how large a price had to be paid for a misunderstanding of both science and religion: to base religious beliefs on an estimate of what science *cannot* do is as foolhardy as it is blasphemous.

The iconoclastic image of science has, however, other components not ascribable to a misconception of its function. For example, the historian Arnold Toynbee charged science and technology with usurping the place of Christianity as the main source of new cultural symbols. Neo-orthodox theologians have called science the "self-estrangement" of man because it carries him with idolatrous zeal along a dimension where no ultimate—that is, religious—concerns prevail. Similarly, T. S. Eliot proclaimed that culture and religion are "different aspects of the same thing," and defined culture and religion in such a way that science, when mentioned at all, would become identifiable with idolatry.[1] It is evident that these views fail to recognize the multitude of divergent influences that shape a culture, or a person. And on the other hand there is, of course, a group of scientists, though not a large one, which really does support the view that science is an iconoclastic activity. Ideologically they are descendants of Lucretius, who wrote on the first pages of *De Rerum Natura*, "The terror and darkness of mind must be dispelled not by the rays of the sun and glittering shafts of day, but by the aspect and the law of nature; whose first principle we shall begin by thus stating, nothing is ever gotten out of nothing by divine power."

ETHICAL PERVERSION

The third image of science is that of a force which can invade, possess, pervert, and destroy a person. The current stereotype of the soulless, evil scientist is the psychopathic investigator in science fiction, or the nuclear destroyer—immoral if he develops the weapons the government asks to be produced, almost traitorous if he refuses that request. According to this view, scientific morality is inherently negative. It causes the arts to languish, it blights culture, and when applied to human affairs, it leads to regimentation and to the impoverishment of life. Science is the serpent seducing us into eating the fruits of the tree of knowledge—thereby dooming us.

The fear behind this attitude is not confined to science; it is directed against all thinkers and innovators. Society has always found it hard to deal with creativity, innovation, and new knowledge. And since science assures a particularly rapid, and therefore particularly disturbing, turnover of ideas, it remains a prime target of suspicion. Factors peculiar to our time intensify this suspicion. The discoveries of "pure" science often lend themselves readily to widespread exploitation through technology. The products of technology—whether they are better vaccines, better gadgets, or better weapons—have the characteristics of frequently being very effective, easily made in large quantities, easily distributed, and very appealing to the masses. Thus we are in an inescapable dilemma—irresistibly tempted to reach for the fruits of science, yet, deep inside, aware that our metabolism may not be able to cope with this ever increasing appetite.

The fear that this dilemma can no longer be resolved increases the anxiety and confusion concerning science. A symptom is the popular identification of science with the technology of weapons. Efforts to convince people that science offers us chiefly knowledge about ourselves and our environment, and occasionally a choice of action, have been rather ineffective. On their side, scientists *as scientists* feel they can take little credit or responsibility either for the facts they discover—for they did not create them—or for the uses others make of their discoveries, for they generally are neither permitted nor specially fitted to make these decisions. Those are controlled by considerations of ethics, economics, or politics, and therefore are shaped by the values and historical circumstances of the whole society. It is, however, also appropriate to say that there has been at least a moderate success in persuading the average scientists, engaged in basic research, of the proposition that the privilege of pursuing a field of knowledge with few restraints imposes on them, in their capacity as citizens, a proportionately larger burden of civic responsibility.

There are other evidences of the widespread notion that science itself cannot contribute positively to culture. Toynbee, for example, gave a long list of "creative individuals," from Xenophon to Hindenburg and from Dante to Lenin, in which he did not include a single scientist. In casual conversation among educated people, it is not fashionable to confess to a lack of acquaintance with the latest ephemera in literature or the arts; but one may even exhibit a touch of pride in professing ignorance of science, of the structure of the universe or one's own body, of the behavior of matter or one's own mind.

THE SORCERER'S APPRENTICE

The last two views hold that man is inherently good and science evil. The next image is based on the opposite assumption—that man cannot be trusted with scientific and technical knowledge. We have survived only because we lacked sufficiently destructive weapons; now we can immolate our world. Science, indirectly responsible for this new power, is here considered ethically neutral. But man, like the sorcerer's apprentice, can neither understand this tool nor control it. Unavoidably he will bring catastrophe on himself, partly through his natural sinfulness and partly through his lust for power, of which the pursuit of knowledge is a manifestation. It was in this mood that Pliny deplored the development of projectiles of iron for purposes of war as "the most criminal artifice that has been devised by the human mind; for, as if to bring death upon man with still greater rapidity, we have given wings to iron and taught it to fly. Let us, therefore, acquit Nature of a charge that belongs to man himself."

When a science is viewed in this plane—as a temptation for the mischievous savage—it becomes easy to suggest a moratorium on science, a period of abstinence during which humanity somehow will develop adequate spiritual or social resources for coping with the possibilities of inhuman uses of modern technical results. Here I need point out only the two main misunderstandings implied in recurrent calls for a moratorium.

First, science of course is not an occupation, such as working on an assembly line, that one may pursue or abandon at will. For creative scientists, it is in large part not a matter of free choice what they shall do. Indeed it is erroneous to think of them as advancing toward knowledge; it is, rather, knowledge which advances toward them, grasps them, and overwhelms them. Even the most superficial glance at the life and work of a major scientist would clarify this point. It would be well if in his or her education each person were shown by example that the driving power of creativity is no less strong and sacred for the scientist than for the artist.

The second point can be put equally briefly. In order to survive and to progress, mankind cannot know too much. Salvation can hardly be thought of as the reward for ignorance. Mankind has been given its mind in order that it may find out where it is, what it is, and who it is, and how it may assume the responsibility for itself which is the chief obligation incurred in gaining knowledge. For example, in the search for the causes and prevention of aggression among people and nations, we shall find the natural and social sciences to be main sources of understanding.

ECOLOGICAL DISASTER

A change in the average temperature of a pond or in the salinity of an ocean may shift the ecological balance and cause the death of a large number of plants and animals. The fifth prevalent image of science similarly holds that while neither science nor man may be inherently evil, the rise of science happened, as if by accident, to initiate a change in the balance of beliefs and ideas that now corrodes the only conceivable basis for a stable society. Jacques Maritain held that the "deadly disease" science set off in society is "the denial of eternal truth and absolute values." Václav Havel's remarks (Chapter 1) echoed this fear.

The main events leading to this state are usually presented as follows: the abandonment of geocentric astronomy implied the abandonment of the conception of the earth as the center of creation and of mankind as its ultimate purpose. Then the idea of purposive creation gave way to blind evolution. Space, time, and certainty were shown to have no absolute meaning. All a priori axioms were discovered to be merely arbitrary conveniences. Modern psychology and anthropology led to cultural relativism. Truth itself has been dissolved into probabilistic and indeterministic statements. Drawing upon analogy with the sciences, other scholars have become increasingly relativistic, denying either the necessity or the possibility of postulating immutable verities, and so have undermined the old foundations of moral and social authority on which a stable society must be built.

However, many applications of recent scientific concepts outside science merely reveal an ignorance about science. For example, relativism in nonscientific fields is generally based on far-fetched analogies. Relativity theory, of course, does not find that truth depends on the point of view of the observer but, on the contrary, reformulates the laws of physics so that they hold good for all observers, no matter how they move or where they stand. Its central meaning is that the most valued truths in science are independent of the point of view. Ignorance of science may also be the excuse for adopting rapid changes within science as models for antitraditional attitudes outside science. In a sense, no field of thought is more conservative than science. Each change necessarily encompasses previous knowledge. Science grows like a tree, ring by ring. Einstein did not prove the work of Newton wrong; he provided a larger setting within which some limitations, contradictions, and asymmetries in the earlier physics disappeared.

But the image of science as an ecological disaster can be subjected to a more severe critique. Regardless of science's part in the corrosion of absolute values, have those values really given us a safe anchor? A priori absolutes abound all over the globe in completely contradictory varieties. Most of the horrors of history have been carried out under the banner of some absolutist philosophy, from the Aztec mass sacrifices to the auto-da-fé of the Inquisition, from the massacre of the Huguenots to the Nazi gas chambers. Few would argue that any society of past centuries provided a meaningful and dignified life for more than a small fraction of its members. If, therefore, some of the new philosophies, inspired rightly or wrongly by science, point out that absolutes have a habit of changing in time and of contradicting one another, if they invite a reexamination of the bases of social authority and reject them when those bases prove false (as did the original colonists in North America), then one must not blame a "relativistic" philosophy for bringing out these faults. They were there all the time.

In the search for a new and sounder basis on which to build a stable world, science will be an indispensable partner. We can hope to match the resources and structure of society to the needs and potentialities of people only if we know more about humans. Already science has much to say that is valuable and important about human relationships and problems. By far the largest part of the total research and development effort in science and engineering today is concerned, indirectly or directly, with human needs, relationships, health, and comforts. Because of the large gap between continuing needs and what has been accomplished so far, and also because of the often slanted and unfair distribution of the fruits of advances, giving credit for real achievements in alleviating human misery is often lagging. Here I am thinking, for example, of the role of science in the expansion of the idea of human rights, especially in the disproving of the old pseudoscientific bases for racism, which were exposed as a sham by the work of anthropologists during the twentieth century. One tends to forget that not all the desirable "applications of science" look like VCRs or pills. Insofar as absolutes are to help guide mankind safely on the dangerous journey ahead, they surely should be at least strong enough to stand scrutiny against the background of developing factual knowledge.

SCIENTISM

While the last four images imply a revulsion from science, scientism may be described as an addiction to science. Among the signs of scientism—which afflicts a small part of the general population but also some scientists[2]—are the habit of dividing all thought into two categories, up-to-date scientific knowledge and nonsense; the view that the theoretical sciences and the large laboratory offer the best models for successfully employing the mind or organizing any effort; and the identification of science with technology, to which reference was made above.

One main source for this attitude is evidently the persuasive success of recent technical work. Another resides in the fact that we are passing through a period of fundamental change in the nature of scientific activity—a change triggered by the perfecting and disseminating of the methods of basic research by teams of specialists with widely different training and interests. Until World War II, the typical scientist worked alone or with a few students and colleagues. Today he or she often belongs to a sizable group. Science has increasingly become a large-scale operation with a potential for rapid and worldwide effects. The result is often a splendid advancement in knowledge, but with side effects which are analogous to those of sudden urbanization—a strain on communication, the rise of an administrative bureaucracy, the depersonalization of human relationships.

To a large degree, this may be unavoidable. The new style of doing science will justify itself by the flow of new knowledge and of material benefits. A danger—and this is the point where scientism enters—is that the fascination with the mechanism of this successful enterprise may change the scientists and society around them. For example, the unorthodox, often withdrawn individual, on whom many great scientific advances have depended in the past, does not fit well into the new system. And society will be increasingly faced with the seductive urging of scientism to adopt generally what is regarded—often erroneously—as the pattern of organization of "Big Science."

MAGIC

Few nonscientists would suspect a hoax if it were suddenly announced that a stable chemical element lighter than hydrogen had been synthesized, or that a manned observation platform had been established at the surface of the sun. To most people it appears that science knows no inherent limitations. Thus, the seventh image depicts science as magic, and the scientist as wizard, deus ex machina, or oracle. The attitude toward scientists on this plane ranges from terror to sentimental subservience, depending on what motives one ascribes to them.

THE IMPOTENCE OF THE MODERN INTELLECTUAL

The prevalence of these images of science is a source of the alienation between the scientific and nonscientific elements in our culture and therefore is important business for all of us. We must consider the full implications of the fact that not only the man and woman in the street but almost all of our intellectual and political leaders today know at most very little about science. And here we come to the central point underlying the analysis made above: the chilling realization that our intellectuals, for the first time in history, are losing their hold on an adequate understanding of the world. Adherence to erroneous images would be impossible were those images not anchored in two kinds of ignorance. One kind is ignorance on the basic level, that of *facts*—what biology says about life, what chemistry and physics say about matter, what astronomy says about the development and structure of our galaxy, and so forth. Most nonscientists realize that the old, common-sense foundations of thought about the world of nature have become obsolete. The simple interpretations of solidity, permanence, and reality have been washed away, and they are plunged into what appears to be a nightmarish ocean of four-dimensional continua, probability amplitudes, indeterminacies, and so forth. They know only two things about the basic conceptions of modern science: that they do not understand them, and that they are now so far separated from them that they may never find out their meaning.

On the second level of ignorance, contemporary intellectuals know just as little about the way in which the different sciences fit together in a world picture. They have had to leave behind them, one by one, those great syntheses which used to represent our intellectual and moral home—the worldview of the book of Genesis, of Homer, of Dante, of Milton, of Goethe. All too many find themselves abandoned in a universe which seems a puzzle on either the factual or the philosophical level. Of all the effects of the separation of culture and scientific knowledge, this feeling of bewilderment and basic homelessness is the most terrifying. Here is one reason, it seems to me, for the frequent self-denigration of contemporary intellectuals. Nor are the scientists themselves unaffected, for it has always been, and must always be, the job of the humanist to construct and disseminate a meaningful picture of the world.[3]

To illustrate this point concretely, and show how it is thought by some to apply since the beginning of modern science, we may turn to a long-respected work by E. A. Burtt, a scholar who understood both the science and the philosophy of the sixteenth and seventeenth centuries. The reader is carried along by Burtt's authority and enthusiasm. And then, suddenly, one encounters a passage unlike any other in the book, an anguished cry from the heart:

> It was of the greatest consequence for succeeding thought that now the great Newton's authority was squarely behind that view of the cosmos which saw in man a puny, irrelevant spectator (so far as a being, wholly imprisoned in a dark room, can be called such) of the vast mechanical system whose regular motions according to mechanical principles constituted the world of nature. The gloriously romantic universe of Dante and Milton, that set no bounds to the imagination of man as it played over space and time, had now been swept away. Space was identified with the realm of geometry, time with the continuity of number. The world that people had thought themselves living in—a world rich with color and sound, redolent with fragrance, filled with gladness, love and beauty, speaking everywhere of purposive harmony and creative ideals—was crowded now into minute corners in the brains of scattered organic beings. The really important world outside was a world hard, cold, colorless, silent, and dead; a world of quantity, a world of mathematically computable motions in mechanical regularity. The world of qualities as immediately perceived by man became just a curious and quite

minor effect of that infinite machine beyond. In Newton, the Cartesian metaphysics, ambiguously interpreted and stripped of its distinctive claim for serious philosophical consideration, finally overthrew Aristotelianism and became the predominant world-view of modern times.[4]

For once, the curtain usually covering the dark fears modern science engenders is pulled away. This view of modern man as a "puny, irrelevant spectator" lost in a vast mathematical system—how far this is from the exaltation that Kepler found through scientific discovery: "Now man will at last measure the power of his mind on a true scale, and will realize that God, who founded everything in the world on the norm of quantity, also has endowed man with a mind which can comprehend these norms!" Was not the universe of Dante and Milton so powerful and "gloriously romantic" precisely because it incorporated, and thereby rendered meaningful, the contemporary scientific cosmology alongside the moral and aesthetic conceptions? Leaving aside the question of whether Dante's and Milton's contemporaries by and large were living in a rich and fragrant world of gladness, love, and beauty, it is fair to speculate that if our new cosmos is felt to be cold, inglorious, and unromantic, it is not the new cosmology which may be at fault, but the absence of new Dantes and Miltons.

To take another concrete example, consider the still widely read book by Arthur Koestler entitled *The Sleepwalkers*. In it, Koestler tried to trace the rise of modern physics (and, with it, of modern philosophical thought), stemming from the work of Kepler, Galileo, Newton, and some of their contemporaries. This is indeed a useful task to set oneself. Koestler worked with devotion on his material. And, most important, he was of course the intelligent layman par excellence whom any scientist would be pleased and proud to have as a colleague and fellow student in an evidently earnest search for an understanding of modern science.

And yet, something terrifying happened as Koestler came to the end of his book. He had been able to see meaning and order in the physics of the seventeenth century. When he turned to modern physics in the epilogue, however, all sense of understanding and coherence disappeared, and the incomprehensible modern conceptions seemed to rise around him on every side as threats to his sanity. As he summarized his work, he found that to a large degree "the story outlined in this book will be recognized as a story of the splitting-off, and subsequent isolated development, of various branches of knowledge and endeavour—sky-geometry, terrestrial physics, Platonic, and scholastic theology—each leading to rigid orthodoxies, one-sided specializations, collective obsessions, whose mutual incompatibility was reflected in the symptoms of double-think and 'controlled schizophrenia.' "[5]

I believe it is important to consider this case as sympathetically as we can—to listen to the anguish of an intelligent man who has discovered that he cannot cope with the modern conceptions of physical reality. For what he is saying is what most people would say—if they were eloquent enough and interested enough in knowledge to be deeply disturbed by a state of threatening ignorance:

> Each of the 'ultimate' and 'irreducible' primary qualities of the world of physics proved in its turn to be an illusion. The hard atoms of matter went up in fireworks; the concepts of substance, force, of effects determined by causes, and ultimately the very framework of space and time turned out to be as illusory as the "tastes, odours and colours" which Galileo had treated so contemptuously. Each advance in physical theory, with its rich technological harvest, was bought by a loss in intelligibility. . . .
>
> Compared to the modern physicist's picture of the world, the Ptolemaic universe of epicycles and crystal spheres was a model of sanity. The chair on which I sit seems a hard fact, but I know that I sit on a nearly perfect vacuum. . . . A room with a few specks of dust floating in the air is overcrowded compared to the emptiness which I call a chair and on which my fundaments rest. . . .
>
> The list of these paradoxa could be continued indefinitely; in fact the new quantum-mechanics consist of nothing but paradoxa, for it has become an accepted truism among physicists that the subatomic structure of any object, including the chair I sit on, cannot be fitted into a framework of space and time. Words like 'substance' or 'matter' have become void of meaning, or invested with simultaneous contradictory meanings. . . .
>
> These waves, then, on which I sit, coming out of nothing,

travelling through a non-medium in multidimensional non-space, are the ultimate answer modern physics has to offer to man's question after the nature of reality.[6]

At the very end of the book, in its last, agitated paragraph, I cannot but hear the cry of a drowning man, a cry for help that cannot leave one unconcerned if one believes that science can and must be shown to play a valid, creative part within our culture:

> The muddle of inspiration and delusion, of visionary insight and dogmatic blindness, of millennial obsessions and disciplined double-think, which this narrative has tried to retrace, may serve as a cautionary tale against the *hubris* of science—or rather of the philosophical outlook based on it. The dials on our laboratory panels are turning into another version of the shadows in the cave. Our hypnotic enslavement to the numerical aspects of reality has dulled our perception of non-quantitative moral values; the resultant end-justifies-the-means ethics may be a major factor in our undoing. Conversely, the example of Plato's obsession with perfect spheres, of Aristotle's arrow propelled by the surrounding air, the forty-eight epicycles of Copernicus and his moral cowardice, Tycho's mania of grandeur, Kepler's sunspots, Galileo's confidence tricks, and Descartes' pituitary soul, may have some sobering effect on the worshippers of the new Baal, lording it over the moral vacuum with his electronic brain.[7]

Burtt and Koestler correctly reflect the dilemma continuing to this day—and I could have chosen equally vivid examples from the more recent avalanche of books with the same animus. What their outbursts tell us, in starkest and simplest form, is this: By having let the intellectuals remain in terrified ignorance of modern science, we have forced them into a position of tragic impotence; they are, as it were, blindfolded in a maze through which they feel they cannot traverse. They are caught between their irrepressible desire to understand this universe and, on the other hand, their clearly recognized inability to make any sense out of modern science. The great literary critic and historian Lionel Trilling testified to this problem frankly in a lecture entitled "The Mind in the Modern World."

The operative conceptions [of science] are alien to the mass of educated persons. They generate no cosmic speculation, they do not engage emotion or challenge imagination. Our poets are indifferent to them. . . .

This exclusion of most of us from the mode of thought which is habitually said to be the characteristic achievement of the modern age is bound to be experienced as a wound given to our intellectual self-esteem. About this humiliation we all agree to be silent, but can we doubt that it has its consequences, that it introduced into the life of mind a significant element of dubiety and alienation which must be taken into account in any estimate that is made of the present fortunes of mind?[8]

Once the alienation of today's nonscientific intellectual is understood, the consequence also becomes plain. The intellectual of tomorrow will probably have an even more distorted image and fearful response with regard to science, for there is at present no countercyclical mechanism at work. Hardly anything being done or planned now is adequate to deal with the possibility of a cultural psychosis engendered by the separation of science and the rest of culture.

Better education cannot be the only remedy; the uses and abuses of science also need reexamination. But making science again a part of every intelligent person's educational resource is the minimum requirement—not because science is more important than other fields, but because it is an integral part of a sound contemporary worldview. A plausible program would include thorough work at every level of education—imaginative new programs and curricula; strengthened standards of achievement; extended college work in science, and expansion of opportunity for adult education. Few people have faced the real magnitude of the problem. Moreover, while some time lag between new discoveries and their wider dissemination has always existed, the increase in degree of abstraction, and in tempo, of present-day science, coming precisely at a time of inadequate educational effort, has begun to change a lag into a discontinuity.

This lapse, it must be repeated, is not the fault of the ordinary citizen, who can only take his cue from the intellectuals—the scholars, writers, and teachers who deal professionally in ideas. Every great age has been shaped by intellectuals who would have been horrified by the proposition that cultivated men and women could dispense with a good grasp of the scientific aspect of the contemporary world picture. This tradition is broken; very few are now able to act as informed mediators. To restore science to reciprocal contact with the concerns of most persons—to bring science into an orbit about us instead of letting it escape from our intellectual tradition—that is the challenge which scientists and all other intellectuals must now face.

Chapter 3

"DOING ONE'S DAMNEDEST": THE EVOLUTION OF TRUST IN SCIENTIFIC FINDINGS

To THIS POINT, we have dealt with the forces trying to reshape the scientific profession's external relations, the part it is thought to play in our culture. But over the past four centuries, the concepts, theories, and practice of science have themselves also undergone an evolutionary development that needs to be understood. It will be shown that the scientists' own *methods* for finding trustworthy research results are the result of a historic progression through stages, each with its own pitfalls and limits, and that here too we have no reason to think our ways will be adequate for tomorrow's needs.

Our mental and physical tools have to develop constantly, if only because it usually becomes more difficult to take the next step as a field evolves. There is within science a marked difference from one end of a century to the other in what it takes to trust one's own data or to trust the results of others. The same evolutionary process has external effects—among them how research is trusted by those outside the scientific community. Today's standards have been assembled from a steady stream of historically identifiable intellectual and social inventions; by the same token, we should expect the tools of science in the future to be in many ways quite different from today's. It will be wise to let the historical view prepare the mind for the next phase. To this end, I shall select a few telling episodes that illustrate the *evolution of trust* in scientific research results.

THE AESTHETICS OF NECESSITY

An appropriate beginning is an example taken from the start of the modern period, which is usually associated with the work of Nicolaus Copernicus. Until some years ago, the reigning opinion in the history of science was that the Copernican system was the very model of a new theory arising from a crisis brought about by the accumulation of data contradicting the old and overly complex theory of Ptolemy and his followers. Copernicus's book on the revolutions of the celestial objects, it was said, used more trustworthy observational data, yielded a better theory, and so rescued the practitioners of the time by giving them at last a calculational method of greater accuracy for astronomical predictions.

But this interpretation is now seen by scholars in the field to be an ahistorical imposition of today's criteria of good behavior at the lab bench. On the contrary, Copernicus is chiefly an exemplar of the early introduction into science of essentially *thematic presuppositions*—that is, of deep convictions about nature on which the initial proposal and eventual acceptance of some of the most powerful scientific theories are still based.[1] Even if Copernicus's main goal had been to help the calculators of ephemerides, which was not the case, he gravely disappointed them. They were no better off after the publication of his great work, *De Revolutionibus*, and in fact they continued to use the Ptolemaic system in essentially the form Ptolemy himself had set forth. For example, Copernicus's system of 1543 gave the same large errors for the predicted location of Mars—up to five degrees—as did the ephemerides of Regiomontanes in the 1470s.

Like Ptolemy, Copernicus selected just enough data (among them many with worse errors than he realized) to get his orbits, even bending some of them by a few minutes of arc as needed. But one must remember what he really wanted to achieve. This he made quite plain near the very beginning of his great book: "To perceive nothing less than the design of the universe and the fixed symmetry of its parts."[2] What is more, he wanted to do so by sticking to what he called "the first principle of uniform motion" (that is to say, Aristotelian circular motion), instead of employing nonuniformity and nonconstancies, as the Ptolemaics had allowed. What convinced him to make this the cornerstone of his argument, and eventually persuaded his followers, was that he thereby produced a model of the planetary system in which the relative locations and order of orbits were no longer arbitrary but followed by necessity. In short, Copernicus is a case study of the privileging of an aesthetically based theory—above all, the aesthetics of necessity—and even of the temporary disbelief in "data" that would appear to disprove a favored theory.

That daring approach has remained to this day, at least during the individual scientist's private, early considerations; and with all its dangers, the glorious thing discovered later was that it often works, and works without bending any data. Indeed, as we shall examine in more detailed discussions about the role of thematic presuppositions, it can turn out that when a thematically and intellectually compelling theory is given a chance, better data, gathered with its aid, will eventually reinforce the theory. That is the meaning behind a remark Einstein made before the test of General Relativity: "Now, I am fully satisfied, and I do not doubt any more the correctness of the whole system, may the observation of the eclipse succeed or not. The sense of the thing is too evident." When a discrepancy of up to 10 percent remained between the first set of measured deviations of light and the corresponding calculations based on his theory, he responded, "For the expert this thing is not particularly important, because the main significance of the theory doesn't lie in the verification of little effects, but rather in the great simplification of the theoretical basis of physics as a whole."[3] And even before more data came in, which decreased the discrepancies, other scientists joined Einstein's camp, persuaded, in the words of H. A. Lorentz, that his grand scheme had "the highest degree of aesthetic merit; every lover of the beautiful must wish it to be true."[4] While this kind of procedure in science is of course much more risky for ordinary mortals, what characterizes giants like Copernicus and Einstein most as they struggle with problems too vast to be solved by the standard procedure—induction from good data—is their intuition of where science will go next. Scientific intuition, when it works, is a gift for which Hans Christian Oersted provided us the happy term "anticipatory consonance with nature."[5]

Yet even the giants of science can't count on their sense of "what's right." When Einstein asked himself what caused him to be so obstinately against a belief in the fundamentality of probabilism in physics, which the success of quantum mechanics forced on most others, he admitted to Max Born that he could not provide logical arguments for his conviction, but could only call on his "little finger as witness." By now we know with fair certainty that in this instance the "little finger" test failed.

INCREASING THE PROBABILITY OF REASON'S CLAIM

But I am getting ahead of my story. On the next stage along the line of the evolution of trust in one's results we find Galileo. Galileo's invention amounted to secularizing science, submerging the qualitative in favor of the quantitative as the earmark of truth, and elevating experimental checks from illustrations of the value of a theory to the test of its *probability*. In a famous passage in the *Two New Sciences* of 1638, Galileo's spokesman, Salviati, goes to such lengths to describe one of these experiments in detail—the accelerated motion of a bronze ball down an inclined plane, including his claim to have "repeated it a full hundred times"—that his supposedly skeptical listener Simplicio is made to confess why he trusts Salviati's account: "I would have liked to be present at these experiments; but feeling confidence in the care with which you perform them, and in the fidelity with which you relate them, I am satisfied and accept them as true and valid." To which Salviati quickly and eagerly responds, "Then we can proceed with our discussion."[6]

This is still uncomfortably close to Aristotle's advice, in his *Rhetorica*, on how to persuade one's listener. Proof provided by the speaker is only part of it. Equally important, persuasion "depends on the personal character of the speaker, and on putting the audience in a right frame of mind." A few decades ago, the science historian Alexandre Koyré, unlike the more pliable Simplicio, found Galileo's account of the detailed care he took in conducting his experiments to be so overblown and unconvincing that he questioned whether Galileo had made any experimental checks at all. Instead, he argued, Galileo had presented only *Gedankenexperimente*, or thought experiments; he was not describing real ones at all.[7]

Fortunately for Galileo's reputation, some pages of his experimental tests have been found and analyzed, and they show that he—who once upon a time had been called the Father of Experimentation—did in fact do experiments. But modern analyses of his lab notes show why Galileo's private test results remained unpublished, and why he could not fully trust them to do more than, in his phrase, *"increase the probability"* of "what reason tells me."[8] At that early stage of the evolution of trust, Galileo's private calculations were largely limited to proportions, and his published work was still narrative. Not until three decades later, in Newton's *Principia*, does the parsimonious style and axiomatic presentation, modeled on Euclid's geometry, take over—the sparse style of public science to which we are accustomed today. In Galileo's books there was still no use of algebra. Galileo does not announce his famous law of free fall as we do in elementary physics, $s = \frac{1}{2} gt^2$. Rather, he makes the mathematically equivalent, but seemingly quite mysterious, statement: "So far as I know, no one has yet pointed out that the distances traversed, during equal intervals of time, by a body falling from rest, stand to one another in the same ratio as the odd numbers beginning with unity."[9]

To have put it this way means that what counted most for Galileo was after all not the limited and perhaps rather silly case of a falling stone or a rolling ball, but the demonstration that terrestrial phenomena, of which these are examples, can be explained by the operation of *integers*—just as the Pythagoreans had dreamed (and as quantum physicists have proved for atomic behavior). Galileo, too, was still engaged in a search for cosmic truths, a tendency which, for better or worse, had to be reined in as science evolved further.

WITNESSING AND PEER REVIEW

The next steps in the evolution of trust were two. They came quickly, and both had to do with the ontological status of a "fact." While Galileo and Newton—a much better experimenter—could still regard their activities in the laboratory mostly as a private affair, others from the middle of the seventeenth century on struggled with the recognition that the door of the lab had to be unlocked; that Simplicio, so to speak, had to be invited to be present while the experiment was being performed. Fact had to be democratized. The academies and the Royal Society devoted much of their energies to the public elicitation of facts, the demonstrations of new phenomena carried out before an audience of interested fellow amateurs, acting as witnesses.[10]

This practice can still be found a century and a half later, in the work of Oersted, who is most remembered for his publication of 21 July 1820 describing the discovery of the interaction of current-carrying conductors and magnets.[11] His hasty and perfunctory laboratory notes of findings, which changed all of physical science, would not be accepted now from a beginning student in an introductory physics course. Nor do they fit the new rules of explicitness that lawyers are busily writing today for government agencies to issue and for us to obey. And Oersted, a true Romantic, was still frank and personal in his publications; he did not know, as Louis Pasteur later told his students, that in their publications scientists have to try to make the results "look inevitable."

Few modern researchers are likely to admit, as Oersted gladly did, that he had been completely convinced many years earlier of the existence of the effect he eventually discovered. Oersted had been persuaded of a connection existing between electricity and magnetism by reading Immanuel Kant, who on metaphysical grounds proposed that all the different forces of nature are only different exemplifications of one fundamental force, a *Grundkraft*.

Oersted is also an example of a scientist still on the early rungs of the evolutionary ladder of trustworthy methods, as is demonstrated by his choice of procedure for what we now would call peer review. There being no other physicists of note available in Copenhagen at the time, the peers he selected to vouch for the truthfulness of his report were brought in more for their presumed moral authority than for their scientific acumen: He assured his readers that the experiments he reported were conducted in the presence of his friend Esmarck, the king's minister of justice, Wleugel, knight of the Order of Danneborg and president of the Board of Pilots, and several other gentlemen whose word, one could assume, was to be trusted.[12]

DECIDING WHAT IS A FACT

In the meantime, between Newton and Oersted, another enormously important problem was being wrestled with: how to determine which of all possible demonstrable events are indications of scientifically usable phenomena; which of them are connected to the fixed regularities of nature and which are merely passing phantoms, clouds with ever-changing form never twice the same, and thus reflecting only ephemeral concatenations? We might call it the problem of telling the difference between signal and noise.

The matter can be summarized by contrasting the styles of research of Robert Boyle, of the second half of the seventeenth century, and of Charles DuFay, of the early decades of the eighteenth. Boyle, while still proclaiming the superiority of reason over authority and even over experience, argues forcefully in his "Proemial Essay" for the inclusion in scientific reports of as many readings and as much detail of the experimentation as possible. He sounds quite modern when he favors "information of sense assisted and heightened by instruments" or argues that "artificial and designed experiments are usually more instructive than observations of nature's spontaneous acting." And he tries to respond to the complaint of Francis Bacon: "Nothing duly investigated, nothing verified, nothing counted, weighed or measured is to be found in natural history; and what in observation is loose and vague is in information deceptive and treacherous."

But Boyle doesn't yet know the difference between mere readings and reliable data. His method of measurement is still quite primitive. For example, in his famous experiments on the compressibility of air, he quietly assumes that the tube in which the gas column is being compressed is of uniform diameter, that the mercury in it is sufficiently degassed, and also that he can make, with high accuracy, naked-eye readings of its level with reference to a paper scale pasted on the outside of the glass tube.

Among Boyle's contemporaries there were a number of fellow amateurs of science and instrument makers who in fact dedicated themselves to increasing the range and accuracy of basic measurements. They typified the sort of person who, from the sixteenth century to our day, has reveled in the invention of more and more ingenious devices for measuring time, distance, angle, or mass. Their achievements have made it possible to obtain more accurate and useful values both of derived quantities (e.g., force, pressure, electric charge) and of the physical constants of nature. That quest is itself a rather heroic chapter in the history of science and technology. It recounts such triumphs as the nearly logarithmic rates of *decrease* in the uncertainties of weighing (three orders of magnitude, from about 1550 to 1950); decrease in the error, in seconds per day, of time measurement (eight orders of magnitude in three centuries, from about 1650); and decrease in the error of astronomical angular measure (five orders of magnitude, from 1600 to the 1920s). The same story is told in the nearly logarithmic rate of *increase* of the resolving power of microscopes (from 0.9 to 0.2 microns, between 1840 and 1880) and of the energy reached for initiating elementary particle reactions (ten orders of magnitude, from about 200 KeV in 1930 to over 1,000 TeV in the mid-1980s). Spectacular feats in improved accuracy are being achieved with increasing frequency. For example, the measured comparison of the charge-to-mass ratios of the proton and antiproton are now known to one part in a billion, the result of an increase in accuracy by eight orders of magnitude between 1950 and 1995.[13]

Taking that route toward greater reliability of results was not in the forefront of Boyle's thought. But more significant for us here is that he was still wide open to anything that might happen before his eyes, and was apt to count everything observable as a fact to be used in his research. In his desire to find the gold of trustworthy detail, he was smothered by slag and mere sand. And that is of course one reason why not he but Richard Townley, while reading Boyle's publication, discovered among Boyle's data what we now call Boyle's law.[14]

With Charles DuFay, as the historian of science Lorraine Daston has pointed out, we see a representative of a new stage of development. In his researches on electricity, DuFay alters his experimental conditions constantly, with the aim of isolating the relevant variable. He is interested only in those "facts" that, he says, are "characteristic of a large class of bodies, not of isolated species," and can therefore be organized into some scheme or simple rule. In short, he intuits the modern characteristic of facticity: that fact is consensual, invariant, and universal. And he signals the necessary next step toward the modern base of trust in experimental reports: He takes on a collaborator, not merely to witness but to try to repeat what DuFay has observed. DuFay realizes that experiments are very difficult to do correctly at an early stage of a science. More generally, he insists that phenomena do not assume the status of fact until other investigators go over the same ground, repeat the observations, and give their consent. This practice was a key invention along the path of increasing trust in research results.

THE EMERGENCE OF TEAMS

If we had enough space, we would now look at the next contributions to the development of trust as seen in the work of Oersted's followers, particularly Faraday and Maxwell. But let me go on to one of Maxwell's successors at the Cavendish Laboratory in Cambridge, England, J. J. Thomson, usually identified as the discoverer of the electron in the 1890s, and to his student and successor, Ernest Rutherford. They exemplify something new in the evolving tale of trust within the lab.

With "J.J.," then still in his mid-thirties, as the new professor, the Cavendish Laboratory officially became a graduate school in 1895. Ernest Rutherford was among the first advanced students to arrive at the lab, soon to be followed by others of great talent, including a few guests from the United States. What occurred there was another step toward the modern phase of science: the forging of a group identity among the participant scientists.

Even before J.J. established himself as a major charismatic leader, he had his students assist him in his experimental researches and left it to them to chase down the details once he himself was satisfied he had obtained the right order of magnitude from theory. But in addition to his natural way of mentoring his assistants and being interested in each student's work, he had another, rather English weapon: regular after-noon tea in his room for all researchers. That was an occasion to develop the sense of community among the researchers, who not only shared the same interests but also depended on one another in their work. By and by, there emerged a core research group around the professor, though only rarely was the collaboration with J.J. so close as to merit, in his view, joint authorship of resulting papers.[15]

The operation, though successful, was still a hand-to-mouth affair, financed largely by student fees and scholarships. Only inexpensive materials, such as glassware blown by the students themselves, were likely to be used. The sociology of the modern laboratory was beginning to come into view, but not its economics. However, the combination of the selection of the advanced students and the socialization of the group as a whole around the more and more famous central figure was so effective that it gave rise to a kind of moral imperative, a bonding. These individuals would mind no hardship and, I should imagine, would rather have committed ritual suicide than betray the scientific norms of the group and of the times. They were not separate, career-oriented passers-through but "citizens" of a community who shared a common spirit.

Rutherford built on this system with a vengeance when he started his own operations at McGill, then at Manchester, and finally at the Cavendish Laboratory. Lab work was now planned more carefully, progress was discussed with each student at least weekly, and a list of research tasks that Rutherford thought could be completed with reasonable success was given to each student annually. Almost as a by-product of his campaign to understand the atomic nucleus, Rutherford's approach was an early example of a style that would soon lead to real team research. Each student still had to make the most of his or her own equipment, and each had been selected by Rutherford with an eye to being "first class," willing to work hard as if in a constant race and with only the most minimal funds. But despite the fact that they all labored on different projects that were linked together only in Rutherford's mind, he expected full loyalty to himself and to the laboratory. He was ever present. Some of his students later reported they felt he had adopted them. Not for nothing was he known to his students as "Papa," according to J. C. Crowther. Trust in each other and in the quality of their data was assured as automatically as it would have been in a healthy family or an isolated tribe of hungry hunters.

Still, the organization of the laboratory by no means reflected modern team research, that is, research in which the participants of a large laboratory work together on a shared project. Nor was the Cavendish operation, while magnificent in its production of major results, "modern" in terms of its understanding of the cost to science of inadequate funding. I once asked James Chadwick, who had discovered the neutron in 1932 after a painstakingly long search at the Cavendish with the most primitive equipment, to describe the atmosphere there at the time. He wrote back that his hope had been to build a small accelerator to settle the question of whether the neutron existed. However,

> no suitable transformer was available and, although Rutherford was mildly interested, there was no money to spend on such a wild scheme. I might mention that the research grant was about £2000 a year, little even in those days for the amount of work which had to be supported. I persisted with the idea for a year or two. . . . I had quite inadequate facilities, and no experience in such matters. I wasted my time—but no money.[16]

Wasting a year or two of a scientist like Chadwick for lack of budgeted funds was soon to become unacceptable, and so would a style of research that failed to take full advantage of the combined talents of a group. The earliest exemplar of that new style in physics was the "family" of young researchers that Enrico Fermi educated and assembled around himself with spectacular results from about 1929 to the mid-1930s. Here, again, group loyalty was extraordinary. This being Rome, and Fermi being thought infallible in quantum physics, his students called him "Il Papa." Money was no real problem; with excellent political instinct, Fermi had secured the patronage of the Italian state for his laboratory needs. And with equally good managerial instinct he designed methods for deciding on promising research lines and then pursuing them fiercely, with a division of labor within the group that could be a model even now.

All this is implied even in Fermi's decision to list as authors of his laboratory's publications every one of the various members of that team—at least in physics a hitherto unheard-of public assignment of credit and sharing of responsibility for the results. Consider the paper reporting the startling discovery of the artificial radioactivity produced with beams of slow neutrons, sent to the *Ricerca Scientifica* on 22 October 1934. This paper can be said to have effectively opened the nuclear age by announcing the discovery of resonance and introducing the concept of the moderator for nuclear reactions, and it figured in the award to Fermi of the Nobel Prize for physics four years later.[17]

There were five authors, given in this order: Fermi, Amaldi, Pontecorvo, Rasetti, and Segrè. We know from their various biographical statements that the young associates of Fermi worked closely under his inspiration, and the listing of authors, with the name of Fermi put first, is an indicator of their relationship. But I must add that when I looked in the archives in Pisa at the lab notebooks the group kept during that project, I found it difficult to disentangle from those laconic entries who did what, and when. Fermi's innovations, in that more innocent age, did not go so far as to foresee the kind of self-protective bookkeeping forced upon labs in our time by the expectation that (as in the case of a researcher in the laboratory of David Baltimore) even the Secret Service might be called in by eager congressional investigators who doubt the trustworthiness of reported scientific results.

As we know, the invention of publicly assigning and distributing responsibility in works that have multiple authorship has been successful beyond all reasonable expectations. Records are being broken year by year as larger and larger groups launch themselves against harder and harder problems. A paper in the *Physical Review D* of 1 June 1992 listed some 365 authors from thirty-three institutions on three continents, and on 3 April 1995 the *Physical Review Letters* carried two articles, by different groups, on the observation of the Top Quark, each with a similarly large number of members spread throughout the world.[18] Such indicators signal a kind of phase change in the social matrix of science, to which we shall have to turn in closing. Before that, however, we must at least briefly note another crucial milestone on the road to greater reliability of findings: the adoption of statistical methods for the analysis of data.

THE STATISTICAL TREATMENT OF DATA

An example from the first decade of the twentieth century will underline how recent, from a historical perspective, are the details of data processing. These issues, such as the use of statistics and even of significant figures, are now so common that many assume they have been with us since the beginning of time. For this example I return to Robert A. Millikan, a scientist whose use of data in his unpublished lab notes I have had occasion to write about;[19] I found it a revealing case study of the way a creative researcher often exercises judgments during the nascent, private phase of experimental work that may look impermissibly arbitrary when examined with the benefit of hindsight. This time I want to look briefly at Millikan's publication of February 1910 in the prestigious *Philosophical Magazine*, his earliest "big" paper on the road to what two years later turned into his triumphant "oil drop" method for measuring the charge on the electron, e.[20]

In 1910, at age forty-two, Millikan was still essentially unknown, and he was near despair about his chances of breaking through to scientific prominence. He had no way of knowing that the February 1910 paper, on a "new modification" of a well-established method of using the motion of water droplets in electric fields to find e, was to point him soon in the right direction. But the paper allows us to glimpse how, at that point in early twentieth-century history, a scientist treated his or her data.

In the section entitled "Results," Millikan frankly confesses to having eliminated all observations on seven of the water drops—these, for various reasons, he decided had to be "discarded." A typical comment of his, on three of the drops, was "Although all of these observations gave values of e within 2 percent of the final mean, the uncertainties of the observations were such that I would have discarded them had they not agreed with the results of the other observations, and consequently I felt obliged to discard them as it was."[21] Today one would not treat data thus.

Millikan then presented his results in tables, in each of which he had gathered the data from one of six series of experiments, and gave also the raw calculations they yielded. Almost every detail there is astounding from our present point of view. Each observation carries Millikan's opinion of its likely reliability: "The observations marked with a triple star are those marked 'best' in my notebook. . . . The double-starred observations were marked in my notebook 'very good.' Those marked with single stars were marked 'good' and the others 'fair.' "[22] Correspondingly, he assigns to each of the six series a weight to be used in averaging all results to obtain a final value; no details are given, but by inspection one sees that in Millikan's mind he correlates nine or ten stars in a series with a weight of seven, seven stars with a weight of six, five stars with a weight of four, three with three, and zero with a weight of one.

Within each of the six series in the tables, there are more surprises.

There is little attempt to keep the significant figures straight during calculations. No reason is given why the readings are clumped the way they are, except that each clump corresponds to a set of drops that are guessed to carry the same number of electron charges. To find a value for *e* within a clump of observations, Millikan uses the average of all the individual data for the voltages needed to balance the various drops and of the times of fall when the electric field is off—instead of calculating the charges drop by drop and then applying statistical data reduction for a final result. Throughout, an air of utter self-confidence pervades the paper. But one must add one more fact: Despite all these "peculiarities," the final result obtained with the new method for the charge of the electron was excellent for its day and could not be improved for many years. Oersted's phrase, "the anticipatory consonance with nature," comes again to mind.

Statistics had entered theory in the eighteenth century, and experimental science by the mid-nineteenth century. But it took much longer to get general agreement on the proper use of statistical analysis—e.g., to deal with such bothersome problems as outriders in lab data, and when or how to "reject" observations; for in the real world of the laboratory, unlike its idealization by many nonscientists, one must be prepared to find even that "at some level, things will happen that we cannot understand, and for which we cannot make corrections, and these 'things' will cause data to appear where statistically no data should exist. . . . The moral is, be aware and do not trust statistics in the tails of the distributions."[23] Only in the mid-twentieth century did good practice on those points become general at the level of the ordinary lab worker. When excellent books such as E. Bright Wilson's *Introduction to Scientific Research* (1952)[24] became available, every student could read what "reasonable procedures" for data acquisition and reduction might be, including under extreme or difficult conditions.

"BIG SCIENCE"

But of course the landscape has changed immensely in the decades since then. Our journey brings us now to the latest and most difficult stage, that of the gigantic thought- and work-collectives in science.

The story is told well in Peter Galison's book, *How Experiments End*.[25] Chapter 4 is devoted to the discovery in the 1970s of so-called neutral-current events, such as the observation of neutrino-electron scattering. The detection of these events led to the confirmation of the theory of the unification of the electromagnetic and weak nuclear forces by Glashow, Weinberg, and Salam—a modern triumph in the *Grundkraft* program of Kant and Oersted.

The theoretical questions and fierce debates concerning the possibility of such a unification took an entirely new turn when, at the big accelerator in CERN, the research institution on the French-Swiss border, a photograph of a single electron event was found by a group from Aachen in January 1973. The detecting instrument they used had been given the name Gargamelle, after the mother of Gargantua. It was a monstrous bubble-chamber device, holding twelve cubic meters of liquid propane—a major engineering project of its own, and a symbol of the interpenetration of engineering and science in the twentieth century. Thousands upon thousands of pictures are exposed in the course of an experiment and then are given to the team trained to scan the negatives. In this run, one exposure was recognized as being strange and new by one of the scanners. She passed it into the hands of a research student, who identified the traces as those of electrons. The following day the student carried it up along the complex hierarchy of the big group to the next rung on the ladder, to the deputy group leader. Thinking it of considerable interest, he in turn brought it to the institute director, who later called it a picture-book example of what they had been expecting for months: a candidate for neutrino-electron scattering.

The crucial point now was to assess the background, i.e., the probability of a masking event. The director took the picture to England, to a fellow expert. And so on, and so on. The circle of belief expanded constantly. But still, this was only one picture, a single event, and not even an event one could immediately reproduce. What a huge difference from the days of DuFay and his followers, all the way to Fermi's group! Yet, as Galison writes: "Experienced bubble chamber experimentalists found the Aachen electron particularly compelling. Their specialty was famous for several critical discoveries grounded on a few well-defined instances. The omega-minus [particle] had been accepted on the basis of one picture, as had the cascade zero. Emulsion and cloud chamber groups had also compiled arguments based on such 'golden events,' including the first strange particles, and a host of kaon decays."[26]

Still, too much was at stake to publish immediately. One might like to believe the evidence, but one could not yet believe *in* it, trust it to be reproducible in principle. One now had to calculate the probabilities of all kinds of other reactions that could masquerade as the one assumed to be happening. Background events had to be ruled out; if one of those was what the exposure was about, it wouldn't test the Glashow-Weinberg-Salam theory at all.

There ensued an agonizing set of discussions within the large collective, as drafts of publications were being debated. We see here the next step in the evolution we are tracing: the internalization within a large group of fellow workers of the array of old but public methods of arriving at trustworthy results—debates among fellow specialists, preparation of publications and refereeing, and all the other safeguards of standards of demonstration. It is as if a new, large organism were doing science within its own boundaries. Finally, Galison concludes, "the members of the collaboration persuaded themselves that they were looking at a real effect. So it was that no single argument drove the experiment to completion. . . . It was a community that ultimately assembled the full argument."[27]

The brief publication of the final result in *Physics Letters* announcing the find boiled down years of work into a few sentences and neat graphs—a far cry from the extensive documentation of Kepler and Boyle.[28] A heterogeneous group of subcollectives from Belgium, Britain, France, Germany, Italy, Switzerland, and the United States had decided to stake their reputations on a new kind of physics. And it is worth noting that the new finding had been distributed to the network of fellow specialists long before the mailing of the 19 July 1973 issue of the journal, through prepublication "preprints." Over the past few decades preprints have become in some branches of science by far the preferred method of "publishing" and of keeping up with the "literature"; the physics department library at Harvard alone receives over four thousand preprints per year from more than five hundred universities and other institutions worldwide. These in turn are now made quickly available through an electronically accessible database specially designed for them. Here, too, we catch a glimpse of the immense rate of change in the way science is being done. The next frontier is already in view: A debate is in progress among physicists at large colliders as to whether "outsiders" should have access to as yet unpublished data, even before the preprint stage.[29]

But to return to the CERN result. The story did not quite end there. By November 1973, an American group at Fermilab, the research center in Illinois, had gathered enough evidence that all this work at CERN was a mistake—that no neutral current existed. They were almost ready to announce their finding. But then they found traces of the effect that "would not go away," and in April 1974 they too published their evidence in favor of the neutral current.[30]

One lesson of this story is that just as one can trace the influence of Bach and Beethoven in some of Schoenberg's music of the early twentieth century, so can one find some habits, methods, and standards of the old days in the megateams of modern science. But the new conglomeration is also a transformed entity, and it deals with a transformed science. In many branches, the more difficult it is to find credible access to the phenomenon, the more dependent the scientist is on apparatus built by others, on data gathered by still others, and on calculations carried through by yet others. When one is immersed in a large group, some members of which might change from month to month, it is not easy to know whom to believe, when to believe it, and how well to believe it. And the ground is shifting once again before our eyes. More than nine hundred physicists had signed on to a single experiment on the planned superconducting collider in Texas before its cancellation, and the number of collaborators on the Human Genome Project is greater still.[31]

All these changes raise a vexing question. Where lies the source of trust—for a participant in the research as well as for the reader of the publication—when most individuals within a team of widely diversified competences cannot vouch for every aspect of the published results, when perhaps not even one person in the whole group can be expected to be fully conversant with every element? A bitter rule of thumb has evolved: whereas credit for success tends to be parceled out unequally (those close to the field usually feel they "know" who in a long list of authors had the best ideas or most novel techniques), an embarrassing error or misconduct by even a single participant is likely to bring discredit to all the members of the team. Thus there is now a big incentive to invent ways to distribute partial responsibility early and proportional justice, when necessary, later. For example, some research papers identify the contributions of individual contributors to a complex group effort. But because of the differences in location, or the wide spectrum of subspecialties involved, or the styles of leadership, etc., there is such a variety in the way teams are run that no single set of rules is likely to emerge. We are watching an experiment in self-governance, analogous perhaps in some ways to the transition that took place when the Pilgrims arrived in the New World and set about the task of forming a new society.

DOING ONE'S DAMNEDEST, REVISITED

Still, some fundamental things apply as time goes by. It was my luck to have known, and to have learned my trade under, one of the last of the physicist-philosophers, P. W. Bridgman—a Nobel Prize winner (1946) for establishing the field of experimental high-pressure physics almost single-handedly, and also the father of the movement in philosophy of science known as operationalism. Operationalism centers on the position that the meaning of a concept is in its measurement or other test. In the 1930s Bridgman made a famous, useful, and very operational statement, usually remembered as:

> The scientific method is doing your damnedest, no holds barred.[32]

For him personally, doing one's damnedest meant total absorption, from 8:00 A.M., when he arrived at the lab by bicycle, rain or shine. It meant incredible productivity: Bridgman published some 230 substantial scientific papers that came to seven volumes of his collected scientific works,[33] as well as several books on the philosophy of science. It meant unceasing dedication to the process of getting things clear in his own mind. His style was completely hands-on, every datum taken by him or by his one dedicated long-term assistant. Most apparatus was built with his own hands. Only two or three joint papers (one with a young chemist named James B. Conant). Very few thesis students— only those whom he could not persuade otherwise. Only a few hundred dollars a year needed for materials. No overhead. No paperwork. Paradise!

NEW DEMANDS ON SCIENTISTS

Bridgman's statement is still a great definition of how to get at trustworthy scientific results. But in the intervening decades the milieu has so evolved that we need to supplement that advice, to bring it into our time. At the very least we must add a few words:

> While doing your damnedest, watch how your presuppositions are holding up; make sure you understand the results of other people along the chain on whose work you are relying; and keep in view that more and more of the findings of science, resonating through society, may have additional results far from those sought initially.

Bridgman himself discovered that last point in his own way. He once thought that science is essentially "value free"; but with the rise of Nazism in the 1930s, he saw that German scientists were no longer able to act as free intellectuals, and were in many cases co-opted as servants of a fascist state. Therefore he published a "Manifesto" in 1939, to the effect that he was closing his lab to visiting scientists from totalitarian countries.[34] The act served as an early, widely noted reminder for scientists in the United States that their work has ethical dimensions even beyond what Jacob Bronowski identified as the principle that binds society together, in science as well as outside, the "principle of truthfulness."

Most scientists today face not only new conditions, such as their complex relationships with sophisticated instruments and their collaborators. They also face a different kind of fact, one that increasingly asserts itself: By obtaining the necessary support from government or industry, they have new obligations, not least the responsibility that their results be shareable far beyond their own labs. For reasons having little to do with the relatively rare case of scientific error, misjudgment, or misbehavior, new pressures for accountability are jostling them from all directions. All parties involved are still fumbling around a bit—the National Institutes of Health, the Congress, the universities, the scientific societies. They are struggling to invent new methods for preventing and defending against charges of misconduct, real or imagined—just as, at an earlier stage of evolution, they had to learn about facticity and significant figures.

This new task, too, is one which history has thrown their way. Scientists cannot let others create the new realities for them. The next phase in the continuing evolution and transformation of the methods of science will depend on the actions of today's researchers. At stake are the problem choices for tomorrow's scientists, their rights to respect and support, the attraction of their fields for future scientists—and even the ancient hope that scientific thought itself is an exemplar for attaining trustworthy conclusions.

Chapter 4

IMAGINATION IN SCIENCE

THE ARTS AND the sciences are typically thought to belong to two different worlds, but in some respects they are close cousins. For while the aims, tools, and products differ, the ingenuity and passion behind the two endeavors are similar. There is also a long history of mutual stimulation between art and science, beginning with Pythagoras, who held that both music and the phenomena of nature are governed by relations among whole numbers. And as we shall see, historians of art have given us key insights into problems in the history of science.

But if we wish to watch the imagination of scientists at work, we shall have to catch them unawares. For quite good reasons—to arrive more dispassionately at consensus—modern scientists try to keep their personal struggles out of their published research results and out of their textbooks. It is therefore chiefly through private records and laboratory notebooks that historians of science discover what scientists themselves do not care to reveal. For while logic, experimental skill, and mathematics are constant guides, they are by no means adequate to the task of scientific investigation—otherwise a computer could do original research unaided. When you listen at the keyhole of the laboratory door, you find that the scientist uses a variety of other tools as well. I shall give examples of three such tools, three closely related companions that are rarely acknowledged: *the visual imagination, the metaphoric imagination, and the thematic imagination.* My examples will be drawn mostly from physics, but one could harvest similar instances from other branches.

THE VISUAL IMAGINATION

One must start with the visual imagination, if only because early Western science made its debut through the eye—through watching the puzzling motions of the planets, the wanderers among the fixed stars. So it is not surprising that entities that could be imagined but were hidden from direct visibility were often regarded with great suspicion. For example, the elusive ether seemed a necessary base for understanding the propagation of light, which consists of transverse electromagnetic waves; but in order to replicate the supposed motions within such an ether one had to think up ever more fantastic mechanical models. Finally the physicist Heinrich Hertz cried a halt to it, saying that the mathematical equations describing light are all that we can imagine when we think of the motion of light waves.

Similarly, the ancient notion of the atom as a tiny, uncuttable, discontinuous entity became more and more inadequate as new electrical, chemical, and other properties of matter had to be explained. At the beginning of the twentieth century, scientists such as Ernst Mach threw up their hands at the very idea of the atom; Mach asked sarcastically: "Have you ever *seen* one?"

In fact, it would not have been impossible to have some sort of physics and chemistry without postulating the existence of atoms (as the chemist Wilhem Ostwald did in the first edition of his text, *Allgemeine Chemie*); but it would have been a much more complex, less beautiful science. Luckily, the eye came to the rescue. In 1912, the physicist C. T. R. Wilson put on display a set of photographs (Figure 1) that settled the matter for most people. He had directed a beam of alpha particles from a radioactive source into a cloud chamber, a small glass box containing moist air at a low temperature. Along the path of each of the alpha particles, which are themselves of course invisible, there settled out a streak of fog, a thin cloud. That gave away the path of the alphas, somewhat as vapor trails in the sky are indirect evidence of the flight of an aircraft.

This was spectacular enough; but the real excitement was in the discontinuities of some tracks, their sudden changes of direction (e.g., the one shown in the lower left corner). There the alpha particle had hit something and scattered off in a new direction. In one case, the obstacle with which it had collided—namely a nucleus in one of the gas

Figure 1

molecules—had been given enough of a push to leave its own tiny vapor trail. These pictures are simple, silent, and still; there is no evidence of motion. By itself, each is an inscrutable hieroglyph. But to the properly prepared mind connected to an alert eye, the photographs presented an overwhelming drama—the first, irrefutable evidence of the existence of atomic discreteness far below the level of direct perception. Scattering of particle beams became *the* way to "see" subatomic events.

Among the next generation of tools that enabled such events to be glimpsed was the bubble chamber—similar to a cloud chamber but filled with a liquid. Now the particles' tracks are traced out in a liquid and become visible as rows of tiny bubbles. Figure 2 is a famous example. The photograph looks again more or less like graffiti, but one learns to neglect in this case the unimportant scratches and curlicues and to focus only on five lines. They reveal that a life-cycle story had taken place on this tiny stage. As drawn in Figure 3, which interprets the raw observations in Figure 2, an elementary particle called a pion came into the view field from the bottom edge. It encountered an

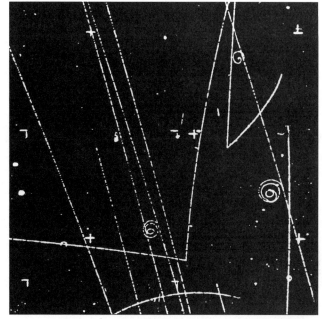

Figure 2

unsuspecting proton in the chamber and interacted with it to form two so-called strange particles (labeled K° and Λ°—"strange" because they survived unexpectedly long for such created particles, namely of the order of one ten-billionth of a second). But these, being neutral, left no traces to look at, until they too decayed. The result of each strange particle's decay was one negative and one positive particle, producing in our view field, as it were, a third generation, each again having its own characteristic lifetime.

You notice that in such a technical description the physicist is using the rhetoric of a familiar, primordial type of drama or folktale, acted out in space and time; a story of birth, adventure, and death. This anticipates a point to be made later: The power of many useful scientific concepts rests at least in part on the fact that they are anthropomorphic projections from the world of human affairs, and to that extent are metaphors.

Next let us look at another bubble-chamber photograph, one first encountered in 1973. By that time the bubble chamber had grown into a monster containing twelve cubic meters of liquid propane and was nicknamed Gargamelle. As noted in Chapter 3, among the thousands upon thousands of pictures taken at CERN, where invisible,

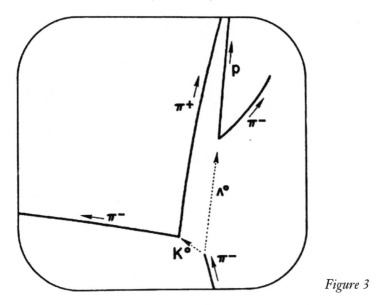

Figure 3

accelerator-created neutrino particles were beamed into the bubble chamber, one of the scanners noticed the frame shown in Figure 4, a photograph unlike any other. On analysis, what she had found turned out to be what is referred to as a "golden event"—a glimpse at a rare but very revealing interaction.

The naive eye must be told to neglect almost everything in that photograph and to focus on the limp curlicue on the left; it is a typical signature of an electron. The interpretation of this event helped finally to confirm the theory of the unification of the electromagnetic and the weak force, called the electroweak force. For that achievement, Sheldon Glashow and Steven Weinberg in the United States and Abdus Salam at Trieste shared a Nobel Prize. I shall come back to this picture shortly to indicate how the scientific imagination dealt with it.

Before we are ready for that exercise, we must go back to the beginning of modern science, to the seventeenth century, to understand better the immense power of the iconic imagination—the ability to form successful mental pictures out of elusive optical images and so to convert vague perceptions into solid knowledge. My example of this process of conversion from optical to mental imagery concerns Galileo Galilei, in a case investigated by the art historian Samuel Edgerton, whose extensive analysis I shall outline.

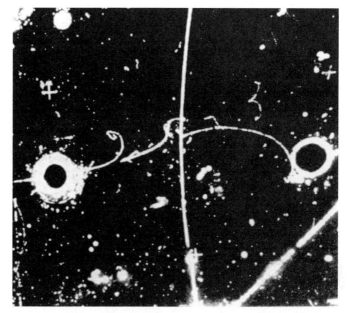

Figure 4

This is the story. In 1609, two men looked at our moon through a new invention, the telescope. The first one to do so was the mathematician, cartographer, and astronomer Thomas Harriot in London, who used a six-powered telescope, starting in late July 1609. The other was Galileo, then a professor of mathematics at the University of Padua; he had taught himself how to grind lenses and made a twenty-powered telescope, with which he observed the moon in late autumn of the same year. Luckily we have a record of what each of these two men thought he saw. It is instructive to compare their private notes, and to understand the reasons for the great differences between them.

Since Aristotle, the moon had widely been thought to be a perfectly smooth, unchanging sphere, the symbol for the incorruptible universe beyond Earth. Also, in paintings since the Middle Ages the moon had been a sign of the Immaculate Conception of the Virgin Mary; Figure 5 (from the painting by Bartolomé Murillo) is an example. But there were two problems with the "perfect" moon. One—of much concern to the religious—was that some areas on the real moon are obviously darker than others, so it couldn't be perfect and uniform throughout. Thomas Harriot called it "that strange spottedness"; and Dante, in canto 2 of the Paradiso section of his *Divine Comedy*, worriedly asked his inspired guide on their journey through the heavens

Figure 5

what the "dusky marks" on the moon portend. (Beatrice put him at ease with an eloquent lecture.)

The second problem was that if the moon were truly a mirror in the shape of a perfect sphere, it would reflect to us at any time the image of the sun on only one small region of it, like a brilliant spot on a ball bearing, leaving the rest dark. But as always, suitable ad hoc theories sprung up to deal with these problems. For example, some said that the moon's surface was ethereal, or like alabaster it reemitted the light in a diffused way, allowing different internal materials to show through.

Harriot's first observation has been preserved among his papers (Figure 6). It is a rough sketch showing the "terminator," the divider between the dark area and the brightly illuminated portion of the moon. But the important thing to note is that evidently Harriot didn't know, and he made no comment on, why it was a *jagged* line instead of

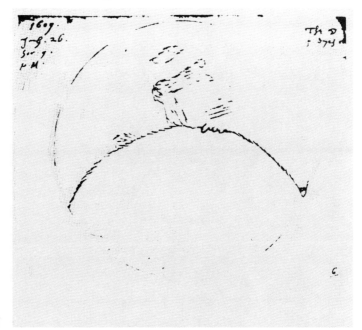

Figure 6

the smoothly curved one which one would expect if the moon were indeed a perfect sphere. He *saw*, but the current theories of the moon's perfection made it difficult for him to understand what he saw.

Enter Galileo. Starting in late November 1609, Galileo carefully follows the moon's phases through his telescope and made several skilled sepia drawings of his observations (Figure 7). Galileo evidently also saw the jagged lines along the terminator. But he interpreted them as irregularities of the surface, as mountains and craters, and he used the chiaroscuro technique of drawing and painting to manipulate dark and light, so as to emphasize the protuberances and depressions.

What Galileo had seen was soon beautifully described in his book of 1610, the *Sidereus Nuncius*. Figure 8 shows one of the illustrations in the book; it exaggerates the moonscape for further effect. Galileo writes that the surface of the moon, contrary to current philosophy, "is not smooth, uniform, and precisely spherical ... but uneven, rough, and full of cavities and prominences, not unlike the face of the earth, relieved by chains of mountains and deep valleys." Galileo *sees* that there is no qualitative difference between earth and moon. He even calculates from the shadows cast by the peaks that the mountains must be four miles high from the baseline—higher than the Alps on earth!

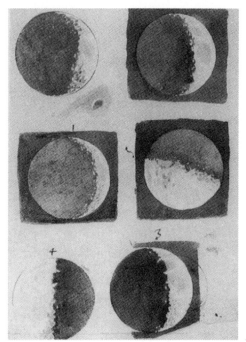

Figure 7

His voice is calm; but he knows that the ancient Aristotelian worldview is crumbling under his blows.

    The news of Galileo's sensational findings spread throughout Europe and transformed what people saw—an example of how the meaning

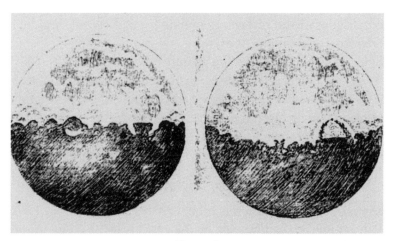

Figure 8

conveyed by objective data depends on the presuppositions held by the viewer. Thomas Harriot himself, having read Galileo's book, raised his telescope again in July 1610, just a year after his first try, and made a sketch of his new observation (Figure 9). Now he too saw mountains and shaded craters—even more of them than were in Galileo's published sketch. Having been converted to a new way of looking, having abandoned his old presuppositions, he now saw something quite different in the same old moon. I am reminded here of that wonderful passage in Tolstoy's *Anna Karenina*, where Anna, hopelessly in love with Count Vronsky, explains to a friend that she cannot love a man like her husband Karenin, because Karenin has such big ears. To this the friend wisely replies that what has changed are not the ears of Anna's husband but the heart of Anna.

Now we must ask what it was that gave Galileo and Harriot such different eyes initially, when they first observed the same object. Part of the answer lies of course in Galileo's greater readiness to consider a Copernican universe, in which planets and satellites can all be similar. Also, by watching the changes in the moon's appearance owing to different illuminations from the sun at different times, Galileo's idea of the analogy between the earth and the moon was bolstered. But much of the answer lies also in their training in visualization, in how they had learned to use their eyes as a tool of imagination. In Harriot's England of 1609, the peak of artistic achievement was the *word*, for example that of Shakespeare, rather than anything in the visual arts. Indeed, visually

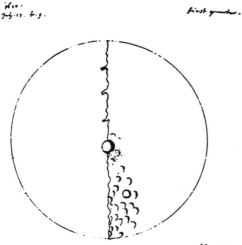

 Figure 9

England was far behind—one might almost say in the Middle Ages—
with respect to understanding perspective renditions. In Galileo's Italy,
however, Renaissance painting had captured the attention of alert
intellectuals. In 1562, under Cosimo I of Florence, Vasari had founded
the great Academy of Design, a center for visual arts and architecture
for the benefit of all, not only the practitioners. It is not an accident that
Galileo's first job application, made at age twenty-five, was for a position as professor of mathematics at the academy, to teach geometry and
perspective. And in 1613 he was elected to membership of that distinguished institution.

So it is very likely that Galileo, like all the students at the Academy
of Design, had studied the problem of how bodies cast shadows on
different surfaces. The typical, well-thumbed texts used in the academy
show how the raised protuberances and depressions on reticulated
spheres appear in light and shade (Figure 10). The art of perspective
rendering in chiaroscuro was a tool and an ability which Galileo had
learned as a young man. It came suddenly to good use when the old
shadow-casting problems reappeared before his eyes in 1609 in the
entirely different context of the telescopic view-field. One may say he
was able to *envisage*, through this still rather poor optical tube, how
scientists everywhere would soon come to see and understand the
phenomena in the solar system.

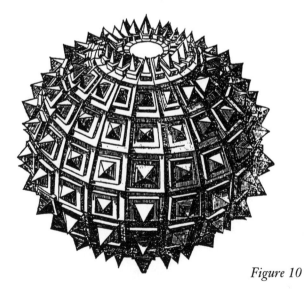

Figure 10

This case was an exemplar of scientific research: hard data plus solid skills of mathematics and praxis, plus theoretical preconceptions, all working together in the theater of the mind. And in this mix, the visual imagination has often been crucial. In a letter to the mathematician Jacques Hadamard, Einstein confessed as follows: "The words or language, as they are written or spoken, do not seem to play any role in my mechanism of thought. The psychical entities which seem to serve as elements in thought are certain signs and more or less clear images which can be voluntarily produced and combined." It was as if, in thought, Einstein played with the pieces of a jigsaw puzzle. And to the psychologist Max Wertheimer, Einstein reported, "I very rarely think in words at all. . . . I have it in a sort of survey, in a way visually." We shall later look more deeply into this fascinating ability. But of course it came in handy when Einstein, as a young man in Bern, became a patent office examiner. His job was to study the descriptions, and particularly the drawings, submitted by inventors, and to reconstruct those proposed machines in his mind to see whether they could function. That task came easily to him. And in his physics, too, he could visualize without effort processes which others found uncomfortably complex.

Let me give you a simple example of visualization. If you have studied physics as far as the introduction of special relativity theory, your textbook no doubt asked you to imagine a train traveling at high speed past a station platform on a stormy day. You were asked to visualize an observer standing on the platform and another seated in the middle of the train. Suddenly two lightning bolts hit the moving train; one strikes in front, and the other one in back. The important question was: How will these events look to each of the two observers, the one who is stationary on the platform and the other traveling at high speed?

You will remember what the answer was: To the first observer the lightning bolts appeared to crash down simultaneously; to the other one (who is moving toward one of the flashes and away from the other), they appeared to have been separated in time. This proves that the simultaneity of two events is not absolute for everyone but depends on the state of motion of the observers; it is "relative" to the framework of the particular observer.

Much of physics follows directly from visualizing this scene in your thoughts, correctly doing this "thought experiment." And that visually dramatic example really does come directly from Einstein's own writings. (A diagram in his 1917 book on relativity contains a typically parsimonious sketch of the situation.) All this was child's play for him, but not so easy for others. It took them time to learn how to see, to imagine it.

Nowadays it has become perhaps too easy. Einstein's imagery has found its way even onto the theater stage. In the five-hour opera by Robert Wilson and Philip Glass titled *Einstein on the Beach*, there is a prominent depiction of that train; but in the opera it is creeping across the stage very slowly for two long acts, and something that represents the lightning bolt is also moving very slowly from above. Einstein would have been amazed to see this, since in his example everything depends on the train moving very fast.

At any rate, Einstein's visual imagination served him superbly again and again. Some time ago I found in the Einstein Archives one of his manuscripts from about 1920, in which he tells how he came to invent the *general* theory of relativity. The key here was to realize that the effects of accelerated motion and of gravity can be considered equivalent. As Einstein described it in the manuscript: One day in 1907 "there came to me the happiest thought of my life," namely that "the gravitational field has only a relative existence. For if one considers an observer who is falling freely from the roof of his house, there exists for him during his fall no gravitational field." For example, any object he releases while falling will fall at the same rate he falls, and therefore, just stay near him. That was a breathtakingly simple, visualizable scientific "thought experiment," the basis of the equivalence principle of general relativity.[1]

In the early part of the twentieth century, the iconic imagination continued to lead from one scientific triumph to another. For example, Niels Bohr's atomic model of 1913 adopted the imagery of the Copernican solar system. It was of course a great breakthrough. But by the mid-1920s it became clear how dangerous it was to think about atomic processes in terms originally invented for large-scale events such as the motion of planets. A new way was needed to imagine phenomena such as the "spin" of the electron, or light being considered both as a wave and a particle. The easily visualized, model-based intuitions, as opposed to conceptual abstraction, had become an obstacle. One does not need to know much about Heisenberg's uncertainty principle to realize that those precisely drawn orbits in Bohr's atomic models cannot exist in nature.

This problem led Heisenberg to propose a necessary but drastic solution, one which to this day makes it difficult for laymen to feel at home in modern physics. Heisenberg totally eliminated the use of picturable models of the atom. A typical Heisenberg dictum was: "The program of quantum mechanics has to free itself first of all from these intuitive pictures. . . . The new theory ought, above all, to give up visualizability totally." Or as the physicist Dirac wrote in 1930: "The classical tradition has been to consider the world to be an association of observable objects. . . . It has become increasingly evident in recent times, however, that nature works on a different plan. Her fundamental laws do not govern the world as it appears in our mental picture in any very direct way, but instead they control a substratum of which we cannot form a mental picture without introducing irrelevancies."

In most other sciences today, the iconic imagination remains alive and well. But the quantum scientists of today have gained a new kind of visualizability, though largely through mathematical rather than physical constructs, through symmetries and abstract diagrams. In Figure 11 we have at least a hint of how the new way of thinking differs from the older one. At the top is pictured the traditional visceral picture one employed to talk about what happens when two equally charged electrons are brought near each other. It is a kind of snapshot of a situation in space; the two electrons mutually exert forces of repulsion across the

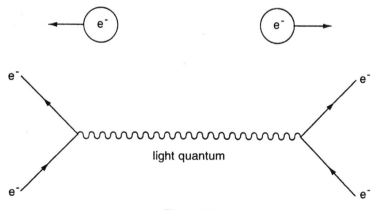

Figure 11

gap between them. But it is now found much more meaningful to think of this phenomenon as caused by the mutual exchange between these two particles of a photon, an entity that mediates the interaction. The lower part of the figure represents this new way of "picturing," by means of a Feynman diagram (named after its inventor, Richard Feynman) that provides a representation in space-time of the scattering of two electrons.

Similarly for the beta decay of the neutron, which was first explained by Enrico Fermi. In the old way of picturing beta decay (top of Figure 12), the interaction between the original neutron and the resulting proton, electron, and neutrino, everything takes place at a single space-time point A. By contrast, and as diagrammed in the lower half of Figure 12, the modern understanding of beta decay imagines the interaction between the particles to be "spread out" in space-time and mediated by the exchange of a W-boson. The two presentations give the same predictions at low energies, but the results are quite different at high energies.

Thus as the simple mental models have withered away, new diagrammatic helpers for our thought processes have taken their place— diagrams in which each part stands for a mathematical expression needed to calculate forces or scattering probabilities. Figure 13 is another example. As my colleague Howard Georgi described it: "An important test of the modern theory of the electroweak force is the existence of neutral currents. This means that weak [i.e., rare, improbable] processes of the type shown can occur, in which an electrically neutral virtual quantum (Z°) is exchanged between a neutrino [represented by the curved line on the left] and a quark [curved line on the right], leaving their identities [e.g., their charges] unchanged."[2]

And this brings us back, as I promised, to the "golden event" I spoke of earlier. For this quote is precisely the description of the underlying text of what happened in the photograph that was given in Figure 4. Our "naked eyes" would see only an unconvincing curlicue in the bubble-chamber picture. But the mind's eye sees, through the use of a Feynman-diagram version of the same phenomenon, that a neutrino scatters an electron without any change of charge; therefore a "neutral current" *exists*; therefore the electroweak theory is right; therefore if you happen to be Glashow, Weinberg, or Salam, who thought of this

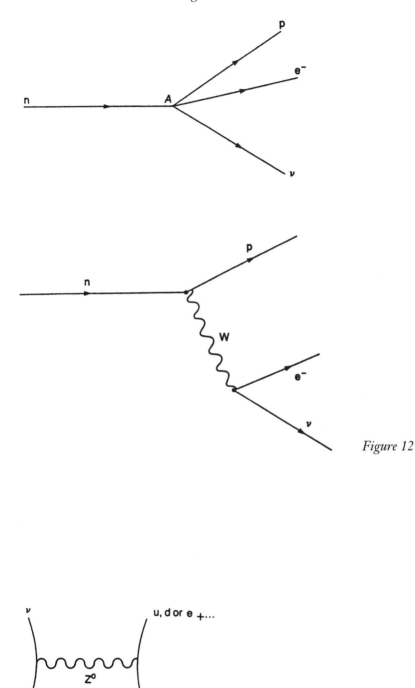

Figure 12

Figure 13

possibility first, you will have reason to be very satisfied with this experimental result.

THE METAPHORIC IMAGINATION

Let us now watch another conceptual tool at work, one that some scientists use with great mastery in the genesis of their ideas. It is the *metaphor* and its close cousin, the *analogy*.

This might surprise you. After all, some philosophers have gone on record to say that the metaphoric imagination has no place in science. The *Dictionary of Modern Thought* still says of metaphor and analogy that they are "a form of reasoning that is particularly liable to yield false conclusions from true premises." Metaphor has been called the "essence of poetry"; it works through illusion. And, surely, the business of scientists is precisely the opposite. Metaphor and analogy might therefore seem to be what scientists should most assiduously *avoid*.

Nevertheless, scientists do use analogies all the time, though they do so quietly. The nineteenth-century physicist Thomas Young is a good example of how one can get chastised if one does it openly. His fame rests chiefly on arguing for the idea that light is fundamentally a wave phenomenon, contrary to the quasi-corpuscular theory that was widely preferred at the time. In one of his first published papers, Young writes: "Light is a propagation of an impulse communicated to [the] aether by luminous bodies." He reminds his reader that "it has already been conjectured by Euler that the colors of light consist of the different frequency of the vibrations of the luminous aether." But this has so far been only a speculation. Now, Young says, he has *confirmation*: The idea that light is a propagation of an impulse in the aether "is strongly confirmed . . ." (by what? how?) "by the *analogy* between the *colors* of a thin plate and the *sounds* of a series of organ pipes" (two entirely different phenomena).[3]

Even without stopping to study the details of this curious and, as it turned out, very fruitful analogy between light and sound—this surprising extension of the concept of wave motion from one field to another, seemingly quite unrelated one—we sense the remarkable daring of this transference of meaning. Indeed, making this connection, and so launching on the experimental proof of the wave nature of light, seemed very ill-advised even to George Peacock, a devoted friend of Young's and himself a scientist at Trinity College, Cambridge. When Peacock published a gathering of Young's articles in 1855—twenty-six years after Young had died, and long after the firm establishment of the wave theory—Peacock still felt he must save his reader from some dreadful mistake on this point; and so he added an asterisk after the crucial sentence just quoted from Young's paper and provided a stern footnote that is perhaps unique in the literature: "This analogy is fanciful and altogether unfounded. Note by the Editor."

Thomas Young's case is an exemplar of the creative but risky function of metaphor or analogy during the nascent phase of the scientific imagination. For Enrico Fermi it was part of his scientific credo to use and reuse the same idea in quite different settings. To him, any physical phenomenon could be understood in terms of an analogy with one of only about a dozen primitive, primary physical situations. For example, he effectively launched modern elementary particle physics with a paper on beta decay in 1934, in which he said the puzzling emission by a nucleus of low-mass particles, such as electrons, should be understood by analogy with the well-established theory of the emission of light quanta (photons) from a decaying atom. In this way he avoided the trap of having to think of the electron as already existing in the nucleus before its emission; after all, no one had felt a need to think of the photon as existing full-blown in the atom before it is radiated away.

And again, soon after writing a paper dealing with the effect that slow *electrons* have on colliding with an atom, Fermi was uniquely able to understand the effect of slow *neutrons* on the nucleus. This happened in October 1934, when he and his team, largely by accident, came upon the miraculously enhanced artificial radioactivity of silver, which turned out to have been caused by scattered, that is, slowed-down, neutrons. The laboratory notebook pages that recorded this discovery are quite laconic, and the resulting paper was very short, less than two pages. But one could say that his use of analogy launched Fermi on what turned out to be the necessary first step toward the nuclear reactor, and hence to the so-called nuclear age.

THE THEMATIC IMAGINATION

With this example, I come to the third type of imaginative tool used by scientists during the nascent phase of research. As noted in Chapter 3, it is the thematic imagination. It is even more risky than those discussed so far: I am referring to the practice of quietly letting a fundamental presupposition—what I have called a *thema*—act for a time as a guide in one's own research when there is not yet good proof for it, and sometimes even in the face of seemingly contrary evidence. This can amount to a willing suspension of disbelief, the very opposite of what one usually takes to be the skeptical scientific attitude.

Indeed, the phrase "willing suspension of disbelief" comes from a discussion of poetry in Samuel Taylor Coleridge's *Biographia Literaria*. He saw his task, as he put it, to imbue his poetic writings with "a semblance of truth sufficient to procure for these shadows of the imagination that *willing suspension of disbelief for the moment* which constitutes poetic faith."

On the face of it, this surely can have nothing whatever to do with science. On the authority of the philosopher of science Karl Popper, we are told that the demarcation criterion of all truly scientific activities is the suspension of belief, not of disbelief. According to Popper, we must subject our rational constructs to a curative purging regime so as to look for a fatal flaw even in our own most treasured brain children. We must try hard to falsify, i.e., disprove them, and therefore to disown them.

And yet, when we stand there at the keyhole of the laboratory door, we observe many of our scientists paying no heed to this well-meant advice. Indeed, sometimes they let their best work grow and mature out of an unlikely idea that they prevent from being destroyed at the hand of iron rationality. Of course, eventually, after this private and nascent phase is over, the results obtained with maturer technique and the guide of maturer theory must stand up to experimental check. Nature cannot be fooled. The graveyard of science is crowded with the victims of some obstinate belief in an idea that proved unworthy. But we must face the strange fact that there *are* genial spirits who can take the risk, and persevere for long periods without the comfort of con-firmatory support, and survive to collect their prizes. By studying their private notes we now know that Isaac Newton, John Dalton, and

Gregor Mendel, among many others, refused to accept "data" that contradicted their thematic presuppositions, and were proven right in the end.

Because this concept is still relatively new and yet crucial for fully understanding how science and scientists progress, we shall further elaborate on it in later chapters (Chapters 5 and 7). But for our purposes here we need only highlight one aspect of the adoption of ardently held themata, and the suspension of disbelief in them: while necessary at some points and often successful, they *can* ultimately mislead terribly. And to conclude with an example of a failure after having spoken of so many scientific successes, let me return to Galileo, and to a long-standing mystery surrounding one of his few but grand errors.

The climax of the scientific revolution for the physical sciences in the seventeenth century was Newton's *Principia*, which combined the imaginative breakthroughs of Galileo Galilei and Johannes Kepler. As Newton confessed, he saw further than others because he had raised himself onto the shoulders of giants such as these. Kepler, at the court of the mad and magnificent emperor Rudolf II in Prague, and Galileo, in Venice and Florence, were two very different personalities; but they had a great deal in common, above all their passionate devotion to the Copernican theory of the planetary system. Each braved the dangers which espousing this heretical notion entailed, and Kepler, the younger by some eight years, an extravagant admirer of Galileo's, tried in every way to get his attention and moral support.

Now it would have been very logical for Galileo to reciprocate, because Kepler's laws clearly showed the superiority of the Copernican way of imagining the system of the world. But contrary to every reasonable expectation, Galileo kept his distance from Kepler, always tried to brush him off, and never accepted Kepler's laws of planetary motion. And that, for a long time, was a maddening puzzle in the history of science. How could Galileo avoid using Kepler's supportive findings as a weapon, when he was so beleaguered by his enemies? What caused this failure of imagination? Galileo never tried to explain his strange rejection of Kepler, and even this shows that it must have had a deep-seated cause. As the historian of science Giorgio de Santillana once said, the ideas of Kepler "must have set in motion a protective mechanism in [Galileo's] mind." What did he want to protect?

The explanation was finally found in a most unexpected way—again by an art historian, the magisterial Erwin Panofsky.[4] His brilliant analysis started from the fact to which I alluded earlier, that Galileo, like so many Italian intellectuals at the time, rightly regarded himself not only as a scientist but also as an admirer and critic of the arts. Moreover, to Galileo, a chief criterion for sound scientific thinking was to use in his science only those thought elements that passed muster on aesthetic grounds. And it was on aesthetic grounds that Galileo found Kepler's ideas unacceptable and even repulsive.

Let me develop Panofsky's argument. Galileo, the son of a re-nowned musician and theorist of music, grew up in a humanistic rather than a scientific environment. It is well known, for example, that he devoted many months of patient labor to a comparison of the poets Ariosto and Tasso, extolling the former and tearing the latter to pieces. Galileo also threw himself gladly into controversies in the visual arts. For example, he was very close to Lodovico Cardi, called Cigoli, considered perhaps the most important Florentine painter of his time. In fact, Cigoli even collaborated with Galileo in astronomical observations; he also called Galileo his "master" in the art of perspective drawing. Cigoli indicated his admiration for Galileo when, in his last work, the frescoes in Santa Maria Maggiore in Rome, he showed the ascending Virgin Mary standing on a moon that matched exactly what Galileo had recorded in one of his illustrations in the *Sidereus Nuncius* (Figures 14 and 15).

In June 1612, Cigoli asked Galileo for help in arguing against those who held that sculpture was superior to painting. Strangely enough, a key to Galileo's rejection of Keplerian astronomy can be found in Galileo's resulting letter on the superiority of painting. The trouble with sculpture, Galileo says, is that it is too closely akin to the "natural things," the objects with which it shares "the quality of three-dimensionality." The painter deserves greater credit for his work pre-cisely because he has available only two dimensions to create the appearance of three-dimensionality. For, as Galileo put it, "The farther removed the *means* of imitation are from the *thing to be imitated*, the more admirable the imitation will be." And he adds for emphasis that we admire a musician if he "moves us to sympathy with a lover by representing his sorrows and passions in song," but not if the musician simply cries and sobs; and we should admire this musician even more if

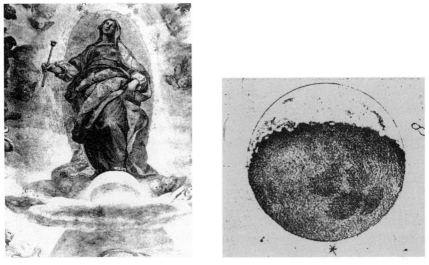

Figure 14   *Figure 15*

he didn't sing at all but used only musical instruments to act on our emotions.[5]

Galileo's point is that we must adhere to "critical purism"; we must keep the *representation* distinct from the subject matter. It is the same sharp knife Galileo used to separate quantity from quality, and science from religion. As Panofsky put it, Galileo objected to any blurring of borderlines. That is precisely why Galileo found Tasso's fanciful allegories so distasteful (for example, in the poem "Gerusalemme Liberata"); and above all why Galileo, like Cigoli, also was opposed to artistic distortions that demeaned the medium of painting, as in "trick pictures." Galileo was particularly scathing of the then widely admired Giuseppe Arcimboldo, court painter to Rudolf II (to make things worse), a painter who had specialized in personifying concepts or seasons by arrangements of implements or of fruits and flowers (see Figure 16, representing summer). This style, now referred to as mannerism, arose as an "anti-classic" tendency, which, Panofsky pointed out, stood opposed "to the ideals of rationality . . . , simplicity, and balance" and instead favored "a taste for the irrational, the fanciful, the complex and the dissonant."[6]

Now there is one element in particular that was as emphatically rejected by high Renaissance art, which Galileo adored, as it was cherished in mannerism, which Galileo abominated. And that is the ellipse. In painting and sculpture the depiction of an ellipse was introduced as a significant element by Correggio and Gian-Maria Falconetto, respectively; in architecture, Michelangelo had briefly toyed with the idea, in designing the shape of the tomb of Pope Julius II, but it was to be only an interior feature, safely invisible from outside. For Galileo, the fight against mannerism, against unnecessary complexity, distortion, imbalance, was a serious duty whether he dealt with music or poetry or painting.

And now we are ready to ask, with Panofsky, "If Galileo's scientific attitude is held to have influenced his aesthetic judgment, might not his aesthetic attitude have influenced his scientific theories?" More specifically, could it be that "both as a scientist and as a critic of the arts [he obeyed] the same controlling tendencies?"[7] We begin to see why Galileo thought Kepler was entirely on the wrong track. On the most obvious level, Kepler's writings, as in his *Mysterium Cosmographicum* and the *Harmonice Mundi*, are such uncontrolled outpourings of different ideas and subjects that it is difficult to see what is valuable under all this seeming fantasy. The three laws of planetary motion of Kepler, without which Newton could not have succeeded, are so deeply buried

Figure 16

under mountains of debris that even Newton found it difficult to acknowledge his indebtedness explicitly.

But quite apart from the indigestibility of Kepler's style of *writing*, Kepler's style of *thinking* seemed to Galileo to enthrone mannerism in the solar system. To Galileo, as to Aristotle and also to Copernicus, all motion in the heavens had to proceed in terms of the superposition of circles, for example in a circular epicycle carried along a circular deferent. The circle, and uniform motion along the circle, were the very signatures of uniformity, perfection, eternity. Kepler had initially thought so, too, but then he had been driven by the data and against his better instincts to announce, as his first law, that the planets are in elliptical motions around the sun. Therefore they were not in what Galileo regarded as "natural" motion but were continually changing their speed as they moved.

To Galileo, who was still completely under the spell of circularity, the ellipse was a distorted circle—a form unworthy of celestial bodies. To accept such an abomination was to give the victory to the Correggios and Arcimboldos of this world. That he would never do. Rather, the primacy of the circle was to Galileo what I have called one of those irresistible thematic presuppositions, without which his scientific imagination could not operate. And not only in the sky but on earth, too. As Galileo put it, "All human or animal movements are circular." Running, jumping, walking, and so on, are, he says, only secondary movements depending on the primary one, which takes place at the joints: "It is from the bending of the leg at the knee, and of the thigh at the hip, which are circular movements, that the jump or the run results."[8]

The enchantment with the circle did not, in the end, undermine Galileo's cosmology in any serious way. But it did harm his physics, for it prevented him from coming to the realization that rectilinear rather than circular movement is the most natural one. Instead, Galileo held—as he put it in book 1 of the *Dialogo*—that Nature allows straight-line motion only temporarily and only for the reestablishment of an order that has been disturbed. Once a piece of matter has reached its proper place, "it has to rest immovable or, if movable, to move only circularly." So Galileo missed the insight that is the very basis of modern mechanics and that we now refer to as Newton's first law, namely, that in the absence of forces, bodies proceed with uniform speed in a straight line. It is truly ironic that the honor of discovering this principle of inertia had to go to the Englishman, who by no stretch of the imagination could have considered himself an admirer or a critic of any of the arts.

We have now seen how some of the best scientists avail themselves of three of the chief tools of the scientific imagination. These examples show again what a caricature is the more common notion of scientific thinking as an almost irresistible, machine-like process of induction. Historians of science and other scholars the world over have been piecing together this more complex and chaotic but more realistic and interesting understanding of how scientists have used their minds while pursuing harder and harder problems during the past four centuries.

But I wish to finish on a cautionary note. By naming the varieties of imagination, we of course shall not have "explained" a Galileo or a Fermi, any more than a Mozart or a Verdi. We shall never fully solve the puzzle of *how* some gifted scientists gain their foreknowledge of the coming state of science, how it is possible for our minds to discover the order of things at all. On this point, Albert Einstein again has to have the last word: "Here lies the sense of wonder, which increases ever more—precisely as the development of knowledge itself increases."[9]

Chapter 5

UNDERSTANDING THE
HISTORY OF SCIENCE

W<small>HAT</small> <small>DOES</small> <small>IT</small> mean to "understand" the history of science—or at least to understand the history of *a* science? To prepare for an answer, let us think of what might be meant by saying that somebody understands a science, for example, physics, and see how that differs from understanding its history. On that question we have some guidance. Albert Einstein, for example, thought that the highest task of a scientist is to achieve an integrated picture of the physical world (a *Weltbild*). This can be understood in terms of an analogy: If one stands on a high mountain, one sees at a glance the varied landscape below, and especially evident are connections in the topography that would be hidden if one looked about from within one of the valleys. Similarly, from an integrated scientific view of nature one should be able to deduce in principle, and in this sense "understand," the detailed phenomena of nature.

We are of course a long way from attaining such a synoptic understanding. Still, a few scientists do have a remarkable comprehension of this sort at least in their own branch of science; we have met some examples. Even when a person of that caliber does not know fairly quickly how to solve a particular problem, he or she will make a good start on it—and is likely to go more or less in the same direction as another scientist of that level. This is probably because of four conditions: First, a problem in one of the exact sciences usually turns out to have one right answer. (That is of course very different from the situation in the history of science.) Next, most such problems are understood in about the same way everywhere; only rarely are there long-lived fundamentally different schools of thought in the exact sciences. (Again, this is quite different in the history of science.) Third, most experts in a given field share more or less the same scientific epistemology and ideology. And lastly, the raw material, the database for most problems in an exact science, is or can be made relatively certain, because it can usually be reproduced or at least reexamined at will in a suitably equipped laboratory. (These last two conditions, also, do not apply to the history of science. For example, a documentary base is much less easily shareable.)

So the first condition for understanding the history of science—and I shall include in it the history of mathematics, technology, and medicine—is that there is little general agreement on procedures or answers. Instead, there are markedly different, competing schools of thought. This has been true since the early period in the field. For example, George Sarton and his followers in the first decades of the twentieth century believed that the main task is the encyclopedic accumulation of precise knowledge of all innovators, throughout all history, and throughout all branches of science and technology. Joseph Needham and his school, on the other hand, approached the history of science through the development of science and technology of one great civilization (China), from its earliest beginnings. Herbert Butterfield felt that the important task was to understand the history of one great period—the rise of science in the seventeenth century. To others, the history of science is chiefly a tool for supplying the "proof" of an ideological or epistemological position (e.g., to find demarcations between "progressive" and "degenerating" research programs). Still others use the history of science to illustrate their theories of the sociology of science (e.g., social construction), or to develop an evolutionary model of scientific progress; and yet another school of thought uses—often the same—cases to support a revolutionary model and the impossibility of scientific progress.[1]

Against this background of competing positions, I shall summarize my own point of view. The history of science is a vigorous and rapidly rising discipline, but unlike the exact sciences it does not have a well-developed theory. I am not at all convinced that any of the theories of history can be applied to the history of science. From the perspective of science itself, the history of science is a young field of scholarship, still in a pre-Newtonian, largely inductive stage. But that does not have to be a handicap for understanding the history of science in a profound way. To that end, we need first of all to do two things: (a) to "understand" thoroughly many of the individual main *events* in the history of a science (events, not speculative constructs such as the "Einsteinian Revolution" or the "synthesis"); and (b) to see *connections* between many of these main events.

What is meant by "event"? Examples of events in the history of science would include the preparation or publication of a scientific paper, or the delivery of an influential speech (such as Niels Bohr's first pronouncement on complementarity at Como in 1927) and transcripts of discussions in the heat of battle, or the documented assurance of a specific discovery (such as the discovery of artificial radioactivity with slow neutrons on a certain day in October 1934 in Fermi's laboratory), or the preparation of an instrument, a letter, or a laboratory photograph (such as C. T. R. Wilson's in 1912), or a page of theoretical manuscript or a laboratory notebook (e.g., Millikan's), or the recording of an oral history interview conducted by a well-prepared historian of science. Each artifact is rooted in the act of doing science and is the preferably unself-conscious physical residue of an action at a given time; it is something that can be studied at a later time and may lend itself to a consensus among competent observers and interpreters who come to the case from different directions. An event in the history of science is in this sense analogous to what an elementary particle physicist calls an event, i.e., a trace on a photograph taken of a bubble chamber or spark chamber. The term is also related to the insistent use of the word *event* in Einstein's relativity paper of 1905; mentioning the word eleven times in a few paragraphs, Einstein implies that science should concern itself in the first place not with fundamentally metaphysically laden notions such as "matter" or "forces" but with what are now called intersections of world lines in space-time.

Focusing on the events helps one avoid the mistake of attempting to impose umbrella conceptions on the raw material, far removed from the underlying factual base in the present state of the field. I would rather try to enter into the belief system inductively, from the shareable evidence of events that involved the personal struggle of ideas.

Different scholars will have different preferences for events that interest and excite them. For example, some opt for paradoxes or incongruities surrounding the event. Others may look for evidence of the operation of institutions, or whole professions, or social movements. Others still, afraid of being charged with "psychologizing," make do only with the final printed product of a scientist and avoid documentary evidence of its context or motivation. My own preference has often been to start with an artifactual residue of an individual's speculative attempts to understand a problem during the nascent phase, which Einstein called "the personal struggle." That is the period *before* the new work has come to some fruition, has been rationalized, pruned of all personal content, published, and absorbed in the mainstream of science through the mechanism of justification. This is, however, *after* the earliest and rarely documentable stirrings of the scientist's new idea. Insofar as I can find the necessary evidence, I tend to focus as much on the state of disorder that precedes the neat conclusion as on the conclusion itself. To study and describe the mind of the scientist in its nascent phase, I look for events that might be observed through a keyhole in the door to the laboratory, to use the metaphor of Chapter 4. In this spirit, Peter Medawar, in *The Art of the Soluble*, asked: "What sort of person is a scientist, and what kind of act of reasoning leads to scientific discovery and the enlargement of the understanding?"[2] He found the usual approaches too limited and issued a challenge: "What scientists *do* has never been the subject of scientific, that is, an ethological inquiry. . . . It is no use looking to scientific 'papers,' for they not merely conceal but actively misrepresent the reasoning that goes into the work they describe. . . . Only unstudied evidence will do—and that means listening at a keyhole."[3]

Often the event that attracts my attention is the writing of a private letter that was not meant to be widely shared, for example, young Albert Einstein's letter of March 1905 to his friend Conrad Habicht. In it, Einstein says he is working on several papers at the same time. The first paper, he says, concerns light radiation, and he calls it "very revolutionary" (one of the very few times Einstein used the word, and he is using it in the sense that in his opinion that work is not based on sound principles but only on a "heuristic" point of view). His second paper, he says, is on the size of the atom, determined from the diffusion phenomena. The third, he explains, is what we now remember as the paper on Brownian movement. And then he adds: "The fourth work is still in draft, and is an electrodynamics of moving bodies, using a modification of the teachings on space and time." (Note the word *modification*—announcing a work considered to be within the existing tradition, the very opposite of a revolutionary act.)

This letter became important because it revealed the possibility of a connection among those four papers, each of which in its published form used to be seen as separate and of very different content. It suggested that one must reread those four publications and find the relationships among those seemingly isolated events. I shall come back to this point later.

Now imagine that a particular letter or article or other artifact, indicating a puzzle or an entry into the nascent phase, has captured your attention and has become your chosen "event." You must next proceed to give the proper *account* (in the sense of a descriptive portrait) of the event. To give the account of an event fully means to begin to understand it. "Accounting" does not consist merely in providing a replication of the letter, article, etc. Rather, it consists of finding and providing as many as possible of the separate *main components* that produced the chosen event. (The list of such components, as you will see, is a metaphor not so very different from the list of active forces present in the creation of any work of scholarship, of literature, or of art.) Purely as a mnemonic device, let me represent the event E under study as a point in a plane, within orthogonal coordinates, the horizontal of which indicates time. E takes place at a given time t (or within an interval Δt) (Figure 1).

There are nine components which in principle make up the full account of any event E in the history of a science. It is unlikely that all nine can be described at once or by the same person engaged in the study of this case; but if an event becomes more fully understood by the profession as a whole, it is because more of the components have been brought in to contribute to the full account.

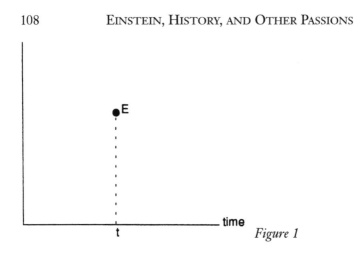

Figure 1

1. *First we must establish an inventory of the contemporaneous state of public, shared scientific knowledge (and misunderstanding or ignorance) of the subject at time* t. What were the published "facts," data, theories, instrumental techniques, and widely believed laws and lore, whether or not they turned out later to have been correct? What did a scientist, or a group of them, believe to be the issue to be faced at a given time, the questions, the tools, indeed the scientific challenges within which this documentary contribution was made? We can call it establishing the historical, publicly available state of science at t (as seen by the scientist in question at that time t).

If we do not do this first, we shall find it almost impossible to avoid the trap of ahistoricity. As a warning, consider the remark of a physicist, writing in a review in the journal *Nature*, that it is an "old myth" that Boltzmann's constant and Avogadro's number were not reliably known before 1905. After all, the author says, Planck's published values in 1900 for these constants, based on his theory of black-body radiation, differed by only 3 percent from today's value. But while this statement is true numerically, it is ahistorical, because in 1905, Planck's theory of radiation of 1900 was still widely ignored. A look at the handbooks of that day shows that these continued to print crude determinations of Avogadro's number while neglecting Planck's values.

2. *Next we need to establish the conceptual development of the public knowledge of a scientific branch, the time trajectory of the state of public*

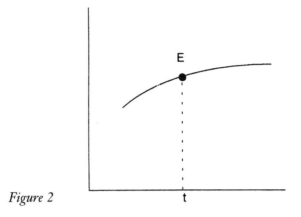

Figure 2

("*shared*") *scientific knowledge leading up to, and if possible going well beyond, the time* t *chosen.* The event *E* is now, as it were, seen to be a point on this line, in a plane in which the horizontal dimension signifies time and the vertical dimension, only quite qualitatively, indicates increasing understanding within a field of science (Figure 2).

This line is merely symbolic of the need to establish the conceptual antecedents. It is what Hans Reichenbach called the "context of justification" within which anything new will have to struggle for its place. Under this heading we would define the tradition or the area of controversy within which a scientist worked and expected his or her students to work. And we should also be able to trace the line beyond *t*, to get closer to the current scientific interpretation of the encounter *E* at *t*. This tracing of "conceptual development" is one of the most frequent activities for historians of science and for historically inclined science educators.

This activity has the largest body of practitioners—it is relatively easy to do—and yet it harbors many dangers. Consider the example where the event *E* stands for the publication of Einstein's relativity paper of 1905.[4] Now the line going through *E* will have points that can be associated with the work of scientists before 1905, and for some years after 1905. Thus we would label one point on the line, corresponding to 1873, as the publication of Maxwell's treatise. Higher on the curve and later in time would be points associated with Heinrich

Hertz's experimental confirmation of Maxwell's electromagnetic theory, Michelson's experiment of 1881, Michelson and Morley's of 1887, and the publications of H. A. Lorentz and Henri Poincaré. And on points beyond 1905, we would place the publication of Walter Kaufmann of 1906, Max Planck's discussion of relativity in 1906, Minkowski's work of 1907, and so on.

This is how textbooks and some history of science courses deal with the development of relativity theory. But one must remember that this scheme is an artificial construction, one that imposes our current view of the "public" science during the period traced by the line. It has many flaws: First, it does not usually go back far enough—to be fair to the case, it should at least reach back to encompass the Galilean-Newtonian view on relativity, as well as Oersted's program of removing the didactic barriers between fields. Second, the usual tracing is in any case only a pedagogic, rationalized device, implying a linear, smoothed-out view of history, with one point or event on this curve seemingly influencing the next one in time causally and necessarily. In actual historic fact, nothing of this sort may have occurred. Rather, the interesting question may sometimes be about the actual discontinuities between parts of the curve rather than the continuities from one point to the next.

But if one bewares the trap of believing such fallacies, a great usefulness emerges from this treatment: One is led to ask what an innovator in fact knew of the related, publicly available developments prior to time t.

3. With this we come to the third component or aspect of an event. What we have done above for the state of public scientific knowledge, we must now do for *the "private" state of scientific knowledge of an individual for whom the event E arises at* t. That is to say, we must now study the less institutional, more ephemeral personal aspects of the scientific activity E at t. We are now looking, to the extent possible, at all material from the workshop and in the desk of a scientist at period t—the object that caught your attention in the first place, the surviving letters, drafts, personal laboratory notes, abandoned equipment, etc. For the sciences more than most other fields, these links to the past are fragile, making it difficult in many cases to provide the "context of discovery." In many of our universities, where every scrap of paper

from literary scholars seems welcome in the archives, it has been more difficult to find a home for scientific correspondence, laboratory records, and apparatus.

Moreover, the need for documentation is not necessarily appreciated or understood by the scientists themselves. Some are still at best impatient with such studies (as with the whole study of what they regard as the "merely personal"). There are evident sociological reasons for that neglect and impatience. The very institutions of science, the selection and training of young scientists, and the internalized image of science are all designed to minimize attention to the personal scientific activity prior to publication. Indeed, the success of science as a shareable activity is connected with the conscious downplaying of the private struggle.

An example of a major scientist reconsidering the value of the past—now happily getting more common—is in the introductory paragraph of Richard P. Feynman's lecture on 11 December 1965, after having been awarded the Nobel Prize the previous day:[5]

> We have a habit in writing articles published in scientific journals to make the work as finished as possible, to cover up all the tracks, to not worry about the blind alleys or to describe how you had the wrong idea first, and so on. So there isn't any place to publish, in a dignified manner, what you actually did in order to get to do the work, although there has been in these days some interest in this kind of thing. Since winning the prize is a personal thing, I thought I could be excused in this particular situation, if I were to talk personally about my relationship to quantum electrodynamics, rather than to discuss the subject itself in a refined and finished fashion. Furthermore, since there are three people who have won the prize in physics, if they are all going to be talking about quantum electrodynamics itself, one might become bored with the subject. So, what I would like to tell you about today is the sequence of events, really the sequence of ideas, which occurred, and by which I finally came out the other end with an unsolved problem for which I ultimately received a prize.

Another obstacle to ready access to documents that might reveal the "private" state of scientific knowledge, particularly in its nascent phase, is the apparent contradiction between the seemingly illogical nature of actual personal discovery and the logical nature of well-developed scientific concepts. This opposition is perceived by some commentators as a threat to the very foundations of science and to rationality itself, whereas in fact one of the secrets of the success of the scientific method is the use of both of these tools during the formative phase of one's research.

In one of his interviews, Einstein urged historians of science to concentrate on comprehending what scientists were aiming at, "how they thought and wrestled with their problems." But he pointed out two difficulties. One was that the scholar would have to have sufficient insight and a kind of empathy (he called it a *"Fingerspitzengefühl"*), both for the content of science and for the process of scientific research, the more so as solid facts about the creative phase are likely to be few. Second, Einstein advised that, as in physics itself, the solution to historical problems may have to come by very "indirect means," and the best outcome to be hoped for is not certainty but only a good probability of being "very likely correct anyway." Historians have come to see the arrogance of claiming that they can reconstruct a historic event "as it really happened." We cannot attain absolute certainty, either in science or in the history of science. But in both cases one may succeed by taking informed, reasonable risks.

4. *Again, as with public science we must next establish the time trajectory of the personal scientific activity under study.* Here we look for the interests and accomplishments and failures of the persons involved over a large portion of their careers. It includes both the period of preparation, on one side of t, and the period of harvesting, on the other side. The work of an individual at t is seen with greater understanding if we have followed the development of his or her personal style prior to t. This second trajectory symbolizes the development of the private or personal science of the person responsible for the event E. Turning again to Einstein, we would identify a point for the year 1894–1895, when he wrote his first scientific essay (on the ether, based on reading Heinrich Hertz). Next we would follow his thought experiment at the age of about sixteen at school in Aarau in 1895–1896, then his reading of Ernst Mach and Hume and others, starting around 1897. Next, perhaps we would consider his confession of "despair" of trying to do

fundamental physics from the experiential base upward, in 1900; and then the work leading to the publication of his various great papers in 1905.

Now we can understand the event E as the intersection of *two trajectories*, one for what we call the private science, a line labeled S_1, and one for the public science, to which trajectory we might give the symbol S_2 (Figure 3).

This point of view opens up immediately some new problems worth working on. One of these would be the acknowledged interaction of S_1 and S_2, with the result that important new names would have to be added to the S_2 trajectory, such as the influence of August Föppl, who has usually been forgotten in the accounts of "conceptual development," but whose book greatly influenced Einstein as a young man. Another problem to consider would be the *lack* of much influence of a work that, while prominent on the profession's S_2 line, happened not to have been absorbed into the individual's S_1 line. This would apply to Lorentz's 1904 formulation of the theory of the electron, which Einstein hadn't read, and the Michelson-Morley experiment, for which Einstein repeatedly, and consistently over many years, specified that it had at most an indirect, and certainly not a crucial, influence on him.

Now let me give an example of the importance of tracing the S_1 trajectory. When you read Einstein's early publications and the letters

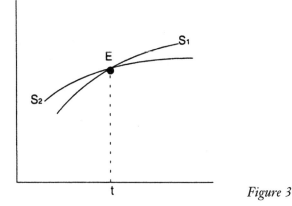

Figure 3

in the archives, you see that the trajectory that brought him to the relativity theory in the miraculous spring of 1905 began in an apparently unpromising way. Einstein's first published article (1901) was entitled "Consequences of the Capillarity Phenomena." For a very young man in need of a job and a career, this was a curious choice. At the time, all the excitement in physics lay in a quite different direction. It was just a few years after the discovery of X rays, radioactivity, and the electron. New experimental findings and new theories chased one another at a dizzying pace. In comparison, capillarity was an old and rather boring subject.

But if we study this paper more carefully, we discover something important. The problem to which Einstein is attending, in this first paper and in the next one, is "the problem of molecular forces." He starts with the promise to proceed from the simplest assumptions about the nature of molecular forces of attraction. In this work, he says, "I shall let myself be guided by analogy with gravitational forces." What interests him, as he writes in a letter to his friend Marcel Grossmann (14 April 1901), is "the question concerning the relationship of molecular forces with Newtonian forces at a distance." Now, that shows great ambition! Einstein's lifelong interest is making its first appearance here: the program of the unification of the various forces of nature. He felt he was working on an important problem. When he writes in that letter that "it is a magnificent feeling to recognize the unity of a complex of phenomena that to direct observation appear to be quite separate things," Einstein is but twenty-two years old; but it is already a familiar Einstein, here searching for bridges between the phenomena of microphysics and macrophysics.

Now let us look again at those major papers of 1905, which Einstein sent off to the *Annalen der Physik* at intervals of less than eight weeks. When I first became interested in this case, it struck me that while the papers—on the quantum theory of light, on Brownian movement, and on relativity theory—seemed to be in quite different fields, the letter to Conrad Habicht I referred to earlier shows that it may have been no accident that they were all done at about the same time. In fact, all can be traced in good part to the same general problem, namely, fluctuation phenomena. Indeed, in the archives I found a letter Einstein wrote to Max von Laue in 1952 in which the connection is indicated.

Einstein discusses there the new edition of von Laue's old textbook on relativity theory and registers an objection:

> When one goes through your account of verifications of the special relativity theory, one gets the impression Maxwell's theory was unchallengeable. But already in 1905 I knew with certainty that Maxwell's theory leads to wrong fluctuations in radiation pressure, and hence to an incorrect Brownian movement of a mirror [suspended in] a Planckian radiation cavity. In my opinion, one cannot get around ascribing to radiation an objective atomistic structure, which of course does not fit into the framework of Maxwell's theory.

Here we see explicitly the chief connection between Einstein's work on Brownian motion of suspended particles, the quantum structure of radiation, and his more general reconsideration of "the electromagnetic foundations of physics" itself. In short, to understand the relativity event E, we need to understand it as part of a more extensive research program.

Moreover, the style of the separate papers, despite their seemingly diverse topics, is essentially the same. Contrary to the sequence one finds in many of the best essays of that time, for example, in H. A. Lorentz's publications on electromagnetic phenomena, Einstein does not start with a review of a puzzle posed by some new and difficult-to-understand experimental fact. Rather, he begins in each case with a statement of his dissatisfaction with what he perceives to be formal asymmetries or other incongruities (which, to other eyes, must have seemed to be predominantly of an aesthetic nature rather than of scientific importance). He then proposes a principle of great generality—one that applies to apparently different parts of physics. Third, he shows that this principle helps remove, as one of the deduced consequences, his initial dissatisfaction. And fourth, at the end of each paper he proposes a small number of predictions that should be experimentally verifiable.

To review: we have traced two lines, two trajectories, one for public science and one for private science; and the event that interests us at a given time begins to be understood as an *occurrence at the intersection of these two trajectories*. We now need to indicate at least briefly the tracing of other strands, the first of which is intertwined with S_1.

5. Here I refer to the *biographical (nonscientific) development* of the scientist under study. The special case of psychobiography is a new and uncertain field, but we have begun to get good material, such as Frank Manuel's and R. S. Westfall's on Isaac Newton, and Erik Erikson's on young Einstein.

6. Another component of a scientific event is its *social setting*, for example, the effect of the educational system on the preparation of scientists in different countries. A case in point is the reception given to the relativity theory in its early years in Germany, Britain, France, and the United States. The differences were enormous during the first decade or two, and a large part of the explanation is found in the differences of the educational systems. For example, in Britain, the preparation of physicists concentrated heavily on the properties of the ether, making it difficult for many of them to break away from that concept. In France, the hierarchical and pyramidal structure of the profession was such that the opposition of Henri Poincaré, the premier scientist in his time, to Einstein's relativity made it very unadvisable for anyone to publish in this field until after Poincaré's death in 1912. In Germany, the fierce competition between autonomous universities (rather than, as in France, the primacy of one university in the country) assured a variety of lively responses. And in the United States, the pragmatic style of understanding the science permitted a surprisingly early acceptance of relativity (though many were under the impression that it was fully supported from the beginning by an experimental base).

Under the same heading of sociological setting, we should also look into the role played in a scientific advance by such effects as that of teamwork in science; the link between academic scientific research and research of interest to industry or the military; and the system for funding research. It would be difficult, for example, to understand the burgeoning of physics research in the United States in the 1920s without paying attention to the role which the grants and policies of private philanthropic foundations such as the Rockefeller and the Guggenheim played at that time. Foundations financed a large fraction of American (and indeed foreign) physicists studying and doing research during the explosive growth of quantum physics. By doing so they helped to create a "critical mass" of scientists whose choice and pace of research were quite different from what they would have been without their grants.

7. A seventh strand deals with the *cultural developments* outside science and the *ideological currents or political events* that influence the work of scientists. Obvious cases spring to mind: the link such scientists as Galileo, Kepler, and Newton saw between science and theology; the link between politics and science in such cases as that of Lysenko, or of the German physicists under the influence of fascism from the 1920s on. But the point is more general, and sometimes the subject of continuing debate. One of these is the case of quantum mechanics; some years ago, a science historian speculated whether developments in the German intellectual environment, including the vogue then enjoyed by Oswald Spengler's book *The Decline of the West* and the rise of revolutionary ideologies, may have prepared some German scientists to abandon the classical principle of causality in the 1920s.

Both this strand and the previously noted influence of the social setting can be so attractive to some students of the history of science that they might be tempted to see all cases as the result of those components alone—an extreme form of the "constructivist" approach. This is as limited a view as its polar opposite—which is narrowly "internalist," focused only on the published product of a scientist's labors.

8. In certain cases the work under study may well be illuminated by an analysis of the philosophical component, particularly of the *epistemological suppositions and the logical structure*. The philosophical worldview of a scientist surely is as important as, for example, his understanding of the mathematical tools of his trade. I am thinking

here of how Einstein confessed to having been influenced by his reading of David Hume and Ernst Mach, or of R. A. Millikan's robust belief in the reality of atoms despite the anti-atomistic teaching of his doctorate supervisor, Michael Pupin, and of others at the time. Or consider that in Germany, and practically nowhere else, the early discussions of quantum mechanics were heavily penetrated by quasi-metaphysical debates on *Anschaulichkeit*—questions on how far scientific concepts can depart from visualizable intuitions—which stemmed chiefly from the hold on German intellectuals of ideas derived from their early reading of Immanuel Kant.

9. So far, we have a list of eight components that in principle can be discerned in the activities from which a given event arises. Last but not least, I turn to the ninth tool for the analysis of a scientific work, which I have termed *thematic analysis*.[6] By this I refer to the often unconfessed or unconscious guiding presupposition a scientist adopts *without* being forced to do so by either data or current theory. An example is Heisenberg's decision to establish a physics built on the thema of discontinuity—thereby freeing physics from what he called, in a letter to Pauli of January 1925, the "swindle" of working with a mixture of quantum rules and classical physics, as Bohr and Sommerfeld still preferred. This decision of Heisenberg's led in turn to his abandoning the customary visualization by which phenomena were thought to be in principle continuous.

Another, and opposite, example of the role of a thematic presupposition is found in Einstein's outburst to Max Born (3 December 1947), briefly mentioned in Chapter 3. Despite the great initial success of quantum theory, Einstein was obstinately opposed to acceptance of the fundamentality of probability. And so he wrote:

> I cannot substantiate my attitude of physics in such a manner that you would find it in any way rational. I see of course that the statistical interpretation . . . has a considerable content of truth, [but] I am absolutely convinced that one will eventually arrive at a theory in which the objects connected by law are not probabilities but facts processed by thought, as one took for granted only a short time ago. However, I cannot provide logical arguments for my conviction, but

can only call on my little finger as a witness, which cannot claim any authority to be respected outside my own skin.[7]

As we shall see in detail, a few such presuppositions guided Einstein throughout his scientific life. The embrace of such themata explains in specific cases why a scientist will try to continue his or her work in a given direction even in the face of ambiguous or contradictory evidence, or why one will refuse to accept theories that are well supported by correlation with phenomena but are based on opposing thematic presuppositions.

As shown by Einstein's reluctance to accept probability, and by Galileo's refusal to abandon the circle (Chapter 4), thematic presuppositions, in themselves not verifiable or falsifiable, can lead one astray if they are held too long against mounting evidence against them. But at least in the nascent phase of scientific work, thematic presuppositions— which Einstein referred to as freely chosen "categories or schemes of thought"—are necessary for most scientists, whether they are used consciously or not. We shall elaborate on these important points in Chapter 7.

The realization of the thematic origins of scientific thought has corrected an appealing but simplistic notion about scientific method that was current in earlier times, and still infects some pedagogic presentations—the notion that the individual scientist always must, and can, start out utterly free from all preconceptions, somewhat as Thomas H. Huxley still thought possible:

> Science seems to me to teach in the highest and strongest manner the great truth which is embodied in the Christian conception of entire surrender to the will of God. Sit down before fact as a little child, be prepared to give up every preconceived notion, follow humbly wherever and to whatever abysses nature leads, or you shall learn nothing. I have only begun to learn content and peace of mind since I have resolved at all risks to do this.[8]

What does save science from falling victim to inappropriate presuppositions are of course the chastening roles both of the coordination with experiment and of the multiple cross-check of any finding by other scientists who may have started with quite different presuppositions. That is why it is now generally thought that Bohr, contrary to Einstein, was right in accepting the thema of fundamental probabilism at the base of atomic theory. Hence, as Dirac pungently put it, "At the present time, one must say that, according to Heisenberg's quantum mechanics, we must accept the Bohr interpretation. Any student who is working for an exam must adopt this interpretation if he is to be successful in his exams."[9] But then Dirac added a warning that one cannot predict which of two antithetical themata will win out in the long run:

> Once he has passed his exams, he may think more freely about it, and then he may be inclined to feel the force of Einstein's argument. . . . It seems clear that the present quantum mechanics is not in its final form. . . . Some day a new quantum mechanics, a relativistic one, will be discovered, in which we will not have these infinities occurring at all. It might very well be that the new quantum mechanics will have determinism in the way that Einstein wanted . . . even though for the time being physicists have to accept the Bohr probability interpretation, especially if they have examinations in front of them.

In sum, the full understanding of a significant event in the history of science requires, to start with, a description and analysis of the event in terms of the nine components laid out here. Since scholarship in the history of science is a communal, collaborative effort, as is science itself, each scholar may be privileged to contribute only a few of these components to a few of the significant events. But in time, the depth of understanding grows by the overlap of these individual contributions. When a historian of science has achieved at least a partial "understanding" of many events ($E_1, E_2, E_3 \ldots$) scattered through time, it becomes possible for that person to achieve an understanding of the field at a second level, in a more profound way. *Connections* between individual events begin to appear, analogous to understanding connecting routes found on a map. One may learn, for example, that Ernest Rutherford and Enrico Fermi, working in different countries on rather different problems, nevertheless influenced each other's work at a crucial period. And of course an event E_1 in the life of a given scientist can have profound and complex relationships with the event E_2 in the same scientist's later work. An instance is how Niels Bohr's papers of 1913 on the structure of atoms already contained the seed of his announcement of complementarity in 1927. Other examples of discovering strands in the dense network of connections that represent the living body of history of science will emerge from our study, focusing chiefly on the publications and influence of Albert Einstein, in what follows.

Part Two

LEARNING
FROM EINSTEIN

Chapter 6

EINSTEIN'S INFLUENCE

ON THE CULTURE OF

OUR TIME

A MAJOR SCIENTIFIC ADVANCE can have a transforming effect on an entire era, on all aspects of a culture. There are excellent studies on the profound influence Newton's triumphs in mechanics and optics had on artists, scholars, and political thinkers for well over a century, ranging from British poets and French philosophers to the eighteenth-century American statesmen whose knowledge of ideas in Newton's *Principia* of 1687 is reflected in the Constitution of the United States. A similar case of far-reaching transformation, felt to this day, is that of James Watt's critical improvement of the steam engine in the 1760s, and its role in helping to launch an industrial revolution which in turn changed the social, economic, intellectual, and ideological bases of life throughout much of the world.

Einstein's scientific publications, especially the early ones on relativity, statistical mechanics, and quantum physics in the first two decades of the twentieth century, also caused remarkable and sometimes quite unforeseen cultural transformations and resonances. As we turn to study this case example, we look first at Einstein's influence on the course of science itself, and then on the rest of the culture of our time.

SCIENCE

Within physics, there was no immediate recognition of the transforming nature of Einstein's early papers. Six years elapsed after the first publication of the special theory of relativity before it had established itself sufficiently to merit a textbook (Max von Laue's *Das Relativitätsprinzip*, 1911), and for some years after that the theory continued to be confused by most scientists with the electrodynamics of H. A. Lorentz. Einstein's ideas on quantum physics, published from 1905 on, were also generally neglected or discounted for years. R. A. Millikan, on accepting his Nobel Prize for 1923, confessed that the validity of Einstein's "bold, not to say reckless" explanation of the photoelectric effect forced itself on him slowly, "after ten years of testing . . . [and] contrary to my own expectation." The transcripts of the questions raised in scientific meetings in the decade after 1905 demonstrated the large intellectual effort required at the time to enter fully into the meaning of the new physics.

But in time, the modern mind opened itself to the counterintuitive ideas. Today, virtually every student who wishes to can learn at least the elements of relativity or quantum physics before leaving high school, and the imprint of Einstein's work on the different areas of physical science is so large and varied that a scientist who tries to trace it would be hard put to know where to start. A modern dictionary of scientific terms contains thirty-five entries bearing his name, from "Einstein: A unit of light energy used in photochemistry" and "Einstein-Bose statistics" to "Einstein tensor" and "Einstein viscosity equation."[1] It is ironic that now, many decades after his death, there is in many branches of the physical sciences more awareness of his generative role than would have been credited during the last twenty years of his life. His ideas became essential for laying out conceptual paths for contemporary work in astronomy or cosmology, in unifying gravitation with the quantum field theory of gauge fields, or even for understanding new observations that were not possible in his time but were predicted by him (as in his 1936 paper which deduced that the gravitational effect of galaxies should act like optical lenses on light).

Apart from changing science itself, Einstein has reached into the daily life of virtually everyone in direct or indirect ways through the incorporation of his ideas on physics into a vast range of technical devices and processes. I need cite only some of the most obvious ones. Every photoelectric cell can be considered one of his intellectual grandchildren. Hence, we are in his debt whenever photo emission or absorption is used, in the home or on the job, to capture an image by means of a television camera, or to project the optical soundtrack of a motion picture, or to set the page of a book or newspaper by photo-composition, or to make a telephone call over fiber cable. In each of these cases, if a law required a label on every appliance giving its intellectual pedigree, such a display would list prominently: "Einstein, *Annalen der Physik* 17 (1905), pp. 132–148; 20 (1906), pp. 199–206," etc.

One would find a label of this sort also on the laser, whose beam was used to lay out the highway on which one travels to the office, or to site the building itself, or to scan the bar code on a store purchase ("Einstein, *Physikalische Zeitschrift* 18 (1917), pp. 121–28," etc.). Or again, the same applies if one lists key ideas that helped to make possible modern electric machinery, such as power generators, or precision clocks that allow the course of planes and ships to be charted. Einstein appears also, if one looks for the ancestry of the ideas, in quantum and statistical physics, by which solid-state devices operate, from calculators and computers to the transistor radio and the car's ignition system; and once more, even when one takes one's vitamin pill or other pharmaceutical drug, for it is likely that its commercial production involved diffusion processes, first explained in Einstein's papers on Brownian movement and statistical mechanics ("Einstein, *Annalen der Physik* 17 (1905), pp. 549–560," etc.).

As Edward M. Purcell has remarked, since the magnetism set up by electric currents is a strictly relativistic effect, derivable from Coulomb's law of electrostatics and the kinematics of relativity, and nothing more, it requires no elaboration to discuss "special relativity in engineering": "This is the way the world *is*. And it does not really take gigavolts or nanoseconds to demonstrate it; stepping on the starter will do it!"[2] It is not much to say that even in our most common experiences, unworldly theoretician's publications help to explain what happens to us all day—indeed, from the moment we open our eyes to the light of the morning, since the act of seeing is initiated by a photochemical reaction ("Einstein, *Annalen der Physik* 37 (1912), pp. 832–838; 38 (1912), pp. 881–884," etc.).

PHILOSOPHY

The proverbial man on the street is quite blissfully ignorant of all that, and has preferred to remain so, even while expecting fully that, mysteriously yet automatically, a stream of practical, benign "spin-offs" continues from the pursuit of pure science. But the philosopher, the writer, the artist, and many others outside the scientific laboratories could not help but be caught up to some extent by the wave that spread beyond science and technology, at first slowly, then with astonishing intensity. As the best scientists were coming to understand what Einstein had done, the trumpets began to sound. Even Max Planck, a person conservative in thought and expression, enthused by 1912: "This new way of thinking . . . well surpasses in daring everything that has been achieved in speculative scientific research, even in the theory of knowledge. . . . This revolution in the physical *Weltanschauung*, brought about by the relativity principle, is to be compared in scope and depth only with that caused by the introduction of the Copernican system of the world."[3] At the same time, on the other end of the philosophical spectrum, the followers and heirs of Ernst Mach rushed to embrace relativity as a model triumph of positivistic philosophy. In the inaugural session of the Gesellschaft für Positivistische Philosophie in Berlin (11 November 1912), relativity was interpreted as an antimetaphysical and instrumentalist conception, and was hailed as "a mighty impulse for the development of the philosophical point of view of our time." When in London on 6 November 1919 the result of the British eclipse expedition was revealed to bear out one of the predictions of general relativity theory, the discussion of implications rose to fever pitch among scholars and laymen, beginning with declarations such as that in *The Times* of London (8 November 1919): the theory had served "to overthrow the certainty of ages, and to require a new philosophy, a philosophy that will sweep away nearly all that has hitherto been accepted as the axiomatic basis of physical thought." It became evident that, as Newton had "demanded the muse" after the *Principia*, now it would be Einstein's turn.

In fact, Einstein did his best to defuse the euphoria and excess of attention that engulfed and puzzled him from that time on. When asked to explain the mass enthusiasm, his answer in 1921 was that "it seemed psychopathological." The essence of the theory was chiefly "the logical simplicity with which it explained apparently conflicting facts in the operation of natural law," freeing science of the burden of "many general assumptions of a complicated nature."[4] That was all. As for being labeled a great revolutionary by his friends and opponents alike, Einstein took every opportunity to disavow the designation. He saw himself essentially as a continuist, had specific ideas on the way scientific theory developed by evolution,[5] and attempted to keep the discussion limited to work done and yet to be done in science. He did not get much help, however. Thus, the physicist J. J. Thomson reported that the archbishop of Canterbury, Randall Davidson, had been told by Lord Haldane "that relativity was going to have a great effect upon theology, and that it was his duty as Head of the English Church to make himself acquainted with it. . . . The Archbishop, who is the most conscientious of men, has procured several books on the subject and has been trying to read them, and they have driven him to what it is not too much to say is a state of *intellectual desperation.*" On Einstein's first visit to England in June 1921, the archbishop of Canterbury therefore sought him out to ask what effect relativity would have on religion. Einstein replied briefly and to the point: "None. Relativity is a purely scientific matter and has nothing to do with religion."[6] But of course this did not dispose of the question. Later that year, even the scientific journal *Nature* felt it necessary to print opposing articles on whether "Einstein's space-time is the death knell of materialism."[7]

Although the crest of the flood, and the worst excesses, have now passed, debates on Einstein's pervasive cultural influence continue. More constructively, since philosophy is concerned in good part with the nature of space and time, causality, and other conceptions to which relativity and quantum physics have contributed, Einstein has had to be dealt with in the pages of philosophers, from Henri Bergson and A. N. Whitehead to the latest issues of the professional journals. As John Passmore observed, it appeared in this century that "physics fell heir to the responsibility of metaphysics."[8]

Some philosophers have confessed that Einstein's work started them off on their speculations in the first place, thus giving some direction to their very careers. One example was Karl Popper, who in his autobiography revealed that his "falsification criterion" (as noted in

Chapter 3) originated in his interpretation of a passage in Einstein's popular exposition of relativity, which Popper said he read with profound effect when he was still in his teens.

Philosophy was no doubt destined to be the most obvious and often the earliest and most appropriate field, outside science itself, to show the influence of Einstein's work. But soon there were others, even though the connections made or asserted were not always valid. From Einstein's wide-ranging output, relativity was invoked most frequently. Cultural anthropology, in Claude Lévi-Strauss's phrase, had evolved the doctrine of cultural relativism "out of a deep feeling of respect toward other cultures than our own"; but this doctrine often invited confusion with physical relativity. Much that has been written on "ethical relativity" and on "relativism" is based on a seductive but misleading play with words. And some art critics have helped to keep alive the idea of a connection between the visual arts and Einstein's 1905 publication.

VISUAL ARTS

Here again, Einstein protested when he could and, as so often, without effect. One art historian submitted to him a draft of an essay entitled "Cubism and the Theory of Relativity," which argued for such a connection—for example, that in both fields "attention was paid to relationships, and allowance was made for the simultaneity of several views."[9] Politely but firmly, Einstein tried to put him straight, and he explained the difference between physical relativity and vulgar relativism so succinctly as to invite an extensive quotation:

> The essence of the theory of relativity has been incorrectly understood in [your paper], granted that this error is suggested by the attempts at popularization of the theory. For the description of a given state of facts one uses almost always only one system of coordinates. The theory says only that the general laws are such that their form does not depend on the choice of the system of coordinates. This logical demand, however, has nothing to do with how the single, specific case is represented. A multiplicity of systems of coordinates is not needed for its representation. It is completely

sufficient to describe the whole mathematically in relation to one system of coordinates.

This is quite different in the case of Picasso's painting, as I do not have to elaborate any further. Whether, in this case, the representation is felt as artistic unity depends, of course, upon the artistic antecedents of the viewer. This new artistic "language" has nothing in common with the Theory of Relativity.[10]

Einstein might well have added here, as he did elsewhere, that the existence of a multiplicity of frames, each one as good as the next for solving some problems in mechanics, went back to the seventeenth century (Galilean relativity). As to the superposition of different aspects of an object on a canvas, that had been done for a long time; thus the eighteenth-century Italian painter Canaletto drew various parts of a set of buildings from different places and merged them in a combined view on the painting (for example, in *Campo S. S. Giovanni e Paolo*).

It was therefore doubly wrong to invoke Einstein as authority in support of the widespread misunderstanding that physical relativity meant that all frameworks, points of view, narrators, fragments of plot, or thematic elements are created equal, that each of the polyphonic reports and contrasting perceptions is as valid or expedient as any other, and that all of these, when piled together or juxtaposed, *Rashomon*-like, somehow constitute the real truth. If anything, twentieth-century relativistic physics has taught the contrary: that under certain conditions we can extract from different reports, or even from the report originating in one frame properly identified, all the laws of physics, each applicable in any framework, each having therefore an invariant meaning, one that does not depend on the accident of which frame one inhabits. It is for this reason that, by comparison with classical physics, modern relativity is simple, universal, and, one may even say, "absolute." The cliché became, erroneously, "everything is relative"; whereas the point is that out of the vast flux one can distill the very opposite: "some things are invariant."

The cost of the terminological confusion has been so great that a brief elaboration on this point will be relevant. Partly because he saw himself as a continuist rather than as an iconoclast, Einstein was reluctant to present this new work as a new *theory*. The term "relativity theory," which made the confusions in the long run more likely, was imposed on Einstein's early work by Planck and Abraham in 1906. For a time Einstein referred to it in print as the "so-called relativity theory," and until 1911 he avoided using the term altogether in the titles of his papers on the subject. In his correspondence Einstein seemed happier with the term *Invariantentheorie*, which is of course much more true to its method and aim. How much nonsense we might have been spared if Einstein had adopted that term, even with all its shortcomings! To a correspondent who suggested just such a change, Einstein replied (letter to E. Zschimmer, 30 September 1921): "Now to the name relativity theory. I admit that it is unfortunate, and has given occasion to philosophical misunderstandings. . . . The description you proposed would perhaps be better; but I believe it would cause confusion to change the generally accepted name after all this time."

LITERATURE

To come back to Einstein's careful disavowal of a substantive link between modern art and relativity: Far from abandoning the quest for it, his correspondent forged onward enthusiastically and published three such essays instead of one. Newton did not always fare better at the hands of eighteenth-century literati and divines who thought they were only following in his footsteps. Poets rush in where scientists fear to tread. And why not, if the apparent promises are so great? In April 1921, at the height of what Einstein on his first journey to the United States all too easily diagnosed as a pathological mass reaction, William Carlos Williams published a poem entitled "St. Francis Einstein of the Daffodils,"[11] containing such lines as "April Einstein / . . . has come among the daffodils / shouting / that flowers and men / were created / relatively equal. . . ." Furthermore, declaring simply that "relativity applies to everything"[12] and that "relativity gives us the clue. . . . So, again, mathematics comes to the rescue of the arts," Williams felt encouraged to adopt a new variable measure for his poems—calling it "a *relatively* stable foot, not a rigid one"[13]—that proved of considerable influence on other poets.

Williams was of course not alone. Robert Frost, Archibald MacLeish, e. e. cummings, Ezra Pound, T. S. Eliot, and some of their disciples (and outside the English-speaking world, others such as

Thomas Mann and Hermann Broch) referred directly to Einstein or to his physics. Some were repelled by the vision thought to be opened by the new science, but there were at least as many who seemed to be in sympathy with Jean-Paul Sartre's remark that "the theory of relativity applies in full to the universe of fiction."[14] Perhaps the most cheerful of the attempts to harness science and literature to common purpose is Lawrence Durrell's entertaining set of novels, *The Alexandria Quartet*, of which its author says by way of preface: "Modern literature offers us no Unities, so I have turned to science and am trying to complete a four-decker novel whose form is based on the relativity proposition. Three sides of space and one of time constitute the soupmix recipe of a continuum."[15] The intention is to use the properties of space and time as determining models for the structure of the book. Durrell says, "The first three parts . . . are to be deployed spatially . . . and are not linked in a serial form. . . . The fourth part alone will represent time and be a true sequel."

For that alone one would not have had to wait for Einstein. But more seems to be hoped for; that, and the level of understanding, is indicated by the sayings of the character Pursewarden in the novel. Pursewarden—meant to be one of the foremost writers in the English language, his death mask destined to be placed near those of Keats and Blake—is quoted as saying, "In the Space and Time marriage we have the greatest Boy meets Girl story of the age. To our great-grandchildren this will be as poetical a union as the ancient Greek marriage of Cupid and Psyche seems to us." Moreover, "the Relativity proposition was directly responsible for abstract painting, atonal music, and formless . . . literature."[16]

Throughout the novel it is evident that Durrell has taken the trouble to read up on relativity, but chiefly out of impressionistic popularizations such as *The Mysterious Universe* by James Jeans, even though Durrell readily confessed that "none of these attempts has been very successful."[17] There is something touching and, from the point of view of an intellectual historian, even a bit tragic about the attempt. In his study *A Key to Modern British Poetry*, Durrell revealed his valid concern to show that as a result of "the far-reaching changes in man's ideas" about the outer and inner universe, "language has undergone a change in order to keep in line with cosmological inquiry (of which it forms a part)."[18] Yet on page after page the author demonstrates that he has been misled by the simplifications of H. V. Routh and James Jeans; he believes that Rutherford and Soddy suggested that the "ultimate laws of nature were not simply causal at all," that "Einstein's theory joined up subject and object," that "sofar as phenomena are concerned . . . the uniformity of nature disappears," and so forth.[19] The terrible but clarifying remark of Wolfgang Pauli comes to mind, who said about a theory that seemed to him doomed: "It is not even wrong."

I have spelled out some of the misunderstandings by which Einstein's work, for better or worse, has been thought to have found its way into twentieth-century culture. But the examples of incorrect interpretation prepare us to appreciate the correct ones that much more. I should confess that my own favorite example of the successful transmutation of scientifically based conceptions in the writer's imagination is to be found in a novel, and a controversial one. William Faulkner's *The Sound and the Fury* is more like an earthquake than a book. Immediately on publication in 1929 it caused universal scandal; for example, not until Judge Curtis Bok's decision in 1949 was this, among Faulkner's other novels, allowed to be sold in Philadelphia.

On the surface it seems unlikely that this book—even a friendly reviewer characterized it as "designedly a silo of compressed sin"—has any resonance with the ideas of modern physics, by intent or otherwise. At the time he poured himself into the book, Faulkner was still almost unknown, largely self-taught, eking out a living as a carpenter, hunter, and coal carrier on the night shift of a power station, using as his desk the upturned wheelbarrow on which he would write while kneeling on the floor. Yet even there he was not isolated, if he read only a small part of the flood of articles in newspapers, periodicals, and popular books in the 1920s that dealt with the heady concepts of relativity theory—such as the time dilation a clock traveling through space undergoes, or the necessity of recognizing the meaninglessness of absolute time and space—and the recent quantum physics, with its denial of the comforts of classical causality. Particularly in America, Einstein was quoted down to the level of local evening papers and *Popular Mechanics*, resulting in wide circulation of such haunting epigrams as his remark, made in exasperation to Max Born (1926), that "God does not throw dice."

Could any of this have reached Faulkner?

In the second of the four chapters of *The Sound and the Fury* we follow Quentin Compson of Jefferson, Mississippi, as he lives through a day in June 1910. It is the end of his freshman year at college and the culmination of a short life wrenched by the degeneration and guilt, the fixations and tribal racism, of his haunted family—from his father, Jason, drinking himself to death, to his feeble-minded brother Benjamin, whose land has been sold to send Quentin to college. The only resource of human affection Quentin has known came from black laborers and servants, although they have been kept in a centuries-old state of terror, ignorance, and obeisance. But the Compsons are doomed. As the day unfolds, Quentin moves toward the suicide he knows he will commit at midnight.

It is all too easy to discover theological and Freudian motives woven into the text, and one must not without provocation drag an author into the physics laboratory for cross-examination when he has already suffered through interrogations at the altar and on the couch. But Faulkner asked for it. Let me select here from a much more extensive body of evidence in the novel itself.

Quentin's last day on earth is a struggle against the flow of time. He attempts to stem the flow, first by deliberately breaking the glass cover of the pocket watch passed down to him from grandfather and father, then twisting off the hands of the watch, and then launching on seemingly random travel, by streetcar and on foot, across the entire city. His odyssey brings him to the shop of an ominous, cyclopean watch repairer. Quentin forbids the man to tell him what time it is, but asks if any of the watches in the shop window "are right." The answer he gets is "No." But wherever he then turns, all day and into the night, he encounters chimes, bells ringing the quarter-hours, a factory whistle, a clock in the Unitarian steeple, the long, mournful sound of a train tracing its trajectory in space and time, "dying away, as though it were running through another month." Even his stomach is a kind of space-time metronome: "The business of eating inside of you space too space and time confused stomach says noon brain says eat o'clock All right I wonder what time it is what of it." Throughout, Quentin carries the blinded watch with him, the watch that never knew how to tell real time and cannot even tell relative time. But it is not dead: "I took out my watch and listened to it clicking away, not knowing it could not even lie."[20] And in the streetcar, the clicking away of time is audible to him only while the car has come to a stop.

Quentin has taken a physics course that freshman year and uses it to calculate the poundage of the weights he needs to buy to drown himself. It is, he says wryly to himself, "the only application of Harvard," and as he reflects on it: "The displacement of water is equal to the something of something. Reductio absurdum of all human experience, and two six-pound flat-irons weigh more than one tailor's goose. What a sinful waste, Dilsey would say. Benjy knew it when Damuddy died. He cried."

As midnight approaches, before he is ready to put his "hand on the light switch" for the last time,[21] Quentin is overcome by torment caused by the shameful memory of his incestuous love for his sister Candace, by her loss, and by his own sense of loss even of the meaningfulness of that double betrayal. In anguish he remembers his father's terrible prediction after he had made his confession:

> You cannot bear to think that some day it will no longer hurt you like this now were getting at it . . . you wont do it under these conditions it will be a gamble and the strange thing is that man who is conceived by accident and whose every breath is a fresh cast with dice already loaded against him will not face that final main which he knows beforehand he has assuredly to face without essaying expedients . . . that would not deceive a child until some day in very disgust he risks everything on a single blind turn of a card no man ever does that under the first fury of despair or remorse or bereavement he does it only when he has realized that even the despair or remorse or bereavement is not particularly important to the dark diceman . . . it is hard believing to think that a love or a sorrow is a bond purchased without design and which matures willynilly and is recalled without warning to be replaced by whatever issue the gods happen to be floating at the time.

This was not the God Newton had given to his time—Newton, of whom, just two centuries before Faulkner's soaring outcry, the poet James Thomson had sung in 1729 that "the heavens are all his own, from the wide role of whirling vortices, and circling spheres, to their great simplicity restored." Nor, of course, was it Einstein's God, a God whose laws of nature are both the testimony of His presence in the universe and the proof of its saving rationality. But this, it seems to me, defines the dilemma precisely. If the poet neither settles for the relief of half-understood analogies nor can advance to an honest understanding of the rational structure of that modern world picture, and if he is sufficiently sensitive to this impotency, he must rage against what is left him: Time and space are then without meaning; so is the journey through them; so is grief itself, when the very gods are playing games of chance, and all the sound and the fury signify nothing. And this leads to recognizing the way out of the dilemma, at least for a few. At best, as in the case of Faulkner, this rage itself creates the energy needed for a grand fusion of the literary imagination with perhaps only dimly perceived scientific ideas. There are writers and artists of such inherent power that the ideas of science they may be using are dissolved, like all other externals, and rearranged in their own glowing alchemical cauldron.

It should not, after all, surprise us; it has always happened this way. Dante and Milton did not use the cosmological ideas of their time as tools to demarcate the allowed outline or content of their imaginative constructs. Those students of ours who, year after year, write dutifully more or less the same essay, explaining the structure of the *Divine Comedy* or *Paradise Lost* by means of astronomy, geography, and the theory of optical phenomena—they may get the small points right, but they miss the big one, which is that the good poet is a poet surely because he can transcend rather than triangulate. In Faulkner, in Eliot's *The Waste Land*, in Woolf's *The Waves*, in Mann's *Magic Mountain*, it is futile to judge whether the traces of modern physics are good physics or bad, for these trace elements have been used in the making of a new alloy. It is one way of understanding Faulkner's remark on accepting his Nobel Prize in 1950: The task was "to make out of the material of the human spirit something which was not there before."[22] And insofar as an author *fails* to produce the feat of novel crystallization, I suspect this lack would not be cured by more lessons on Minkowski's space-time, or Heisenberg's indeterminacy principle, or even thermodynamics, although these lessons could occasionally have a prophylactic effect that might not be without value.

The Process of Transformation

Like gifted practitioners of the arts, gifted scientists do not build by patiently assembling blocks that have been precast by others and certified as sound. On the contrary, they too melt down the ready-made materials of science and recast them in a way that their contemporaries tend to think is outrageous. That is why Einstein's own work took many years to be appreciated even by his most notable fellow physicists. His physics looked to them like alchemy, not because they did not understand it, but because, in one sense, they understood it all too well. From their thematic perspective, Einstein's work was anathema. Declaring, by simple postulation rather than by proof, Galilean relativity to be extended from mechanics to optics and all other branches of physics; dismissing the ether, the playground of most nineteenth-century physicists, in a preemptory half-sentence; depriving time intervals of absolute meaning; and other such outrages, all delivered in a casual, confident way in the first, short paper on relativity—those were violent and "illegitimate" distortions of science to almost every physicist. As for Einstein's new ideas on the quantum physics of light emission, Max Planck felt so embarrassed by it when he had to write a letter of recommendation for Einstein seven years later that he asked that this work be overlooked in judging the otherwise promising young man.

The process of transformation characterizes not only science and the flow of ideas from high science to high literature. It also works across the boundaries in other ways. It seems clear to me that without this process of transformation, willing or unwilling, of ideas from science and from philosophy, physics itself would not have come into its twentieth-century form. The case of Einstein suggests, therefore, that the accomplishments of the major innovators—and not only in science—depend on the ability to persevere in four ways: by being loyal primarily to one's own belief system rather than to the reigning faith; by perceiving and exploiting the man-made nature and plasticity of human conceptions; by demonstrating eventually that the new unity that has been promised can and does become lucid and convincing to lesser mortals active in the same field; and, in those rare cases, even by issuing ideas that lend themselves, quite apart from misuse and oversimplification, to further adaptation and transformation in the imagination of similarly exalted spirits who live on the other side of disciplinary boundaries.

PERSONAL INFLUENCE

It remains to deal with one more, somewhat different mechanism by which Einstein's imprint came to be felt by society, far beyond his own field of primary attention: the power of his personal intervention on behalf of causes ranging from the establishment of a homeland for a persecuted people to his untiring efforts, over four decades, for peace and international security. In retrospect we can see that he had the skill, at strategic periods in history, to lend his ideas and prestige to the necessary work of the time. Even the most famous of these personal interventions, the letter to President Roosevelt in 1939 to initiate a study on whether the laws of nature allow anyone to produce an atomic weapon, was of that sort, although it has perhaps been misunderstood more widely than anything else Einstein did. He was, after all, correct in his perception that the Germans, who were pushing the world into a war, had all the skill and intention needed to start production of such a weapon if it was feasible. In fact, they had a head start, and but for some remarkable blunders and lack of means they might have fulfilled the justified fears, with incalculable consequences for the course of civilization.

To highlight these personal interactions, I select one as more or less exemplary of the considerable effect Einstein had even on those who had merely brief or casual discussions with him. The Swiss psychologist Jean Piaget's work entered its most important phase with the publication in 1946 of *The Child's Conception of Time*. The book begins with a plain acknowledgment: "This work was promoted by a number of questions kindly suggested by Albert Einstein more than fifteen years ago (1928, at a meeting in Davos). . . . Is our intuitive grasp of time primitive or derived? Is it identical with our intuitive grasp of velocity? What if any bearing do these questions have on the genesis and development of the child's conception of time? Every year since then we have made a point of looking into these questions. . . . The results (concerning time) are presented in this volume; those bearing on the child's conception of motion and speed are reserved for a later work."[23] Throughout his later writings, Piaget remarked on this debt: "It was the author of the theory of relativity who suggested to us our work,"[24] or "Einstein once suggested we study the question from the psychological viewpoint and try to discover if there existed an intuition of speed independent of time."[25] In addition, Piaget referred explicitly to notions of relativity and other aspects of Einstein's work.[26] Einstein came to have an immense correspondence with leaders in virtually every type of endeavor, and exerted on many a seminal influence of the type that Piaget acknowledged.[27]

THE SPECTRUM OF RESPONSES

Looking back at the variety of ways in which Einstein came to impress the imagination of his time and ours, we can discern some rough categories in the responses to him, spread out, as it were, in a spectrum from left to right. At the center of the spectrum, corresponding to the largest intensity, one finds a widespread but unfocused and mostly uninformed fascination with Einstein, manifested in a variety of ways, from enthusiastic mass gatherings to glimpse the man, to the outpouring of popularizations with good intentions, to responses that betray the vague discomfort aroused by his ideas. A good example of the last is an editorial entitled "A Mystic Universe" in *The New York Times* of 28 January 1928 (p. 14): "The new physics comes perilously close to proving what most of us cannot believe. . . . Not even the old and much simpler Newtonian physics was comprehensible to the man on the street. To understand the new physics is apparently given only to the highest flight of mathematicians. . . . We cannot grasp it by sequential thinking. We can only hope for dim enlightenment." The editorial writer then notes that the ever-changing scene in physics does offer some "comfort":

> Earnest people who have considered it their duty to keep abreast of science by readapting their lives to the new physics may now safely wait until the results of the new discoveries have been fully tested out by time, harmonized and sifted down to a formula that will hold for a fair term of years. It would be a pity to develop an electronic

marriage morality and find that the universe is after all ether, or to develop a wave code for fathers and children only to have it turn out that the family is determined not by waves but by particles. Arduous enough is the task of trying to understand the new physics, but there is no harm in trying. Reshaping life in accordance with the new physics is no use at all. Much better to wait for the new physics to reshape our lives for us as the Newtonian science did.

Similarly, in Tom Stoppard's play *Jumpers*, a philosopher is heard to ask: "If one can no longer believe that a twelve-inch rule is always a foot long, how can one be sure of relatively less certain propositions?"

Near this position, as we said, are the enthusiastic misapplications, usually achieved by an illicit shortcut of meaning from, say, the true statement that the operational definition of length is "framework"-dependent, to the invalid deduction that mental phenomena in a human observer have *thereby* been introduced into the very definitions of physical science. The irony here is that the first lessons we learned from relativity physics were that short circuits in signification must be avoided, for they were just what encumbered classical physics, and that attention must be paid as never before to the meanings of the terms we use.

When we now glance further toward the left, or blue, end of the spectrum, the expressions of resignation and futility become more explicit. Indeed, among some of the most serious intellectuals there seems to be, on this point, a sense of despair. By the very nature of their deep motivation they must feel most alienated from a universe whose scientific description they can hardly hope to understand except in a superficial way. The much-admired humanistic scholar Lionel Trilling spoke of the "exclusion" of nonscientific intellectuals from "the characteristic achievement of the modern age" as a "wound ... to [their] intellectual self-esteem," as quoted fully in Chapter 2.

Einstein, who had intended originally to become a science teacher, came to understand this syndrome, and the obligation it put on him. He devoted a good deal of time to popularization of his own. His avowed aim was to simplify short of distortion. In addition to a large number of essays and lectures, he wrote, and repeatedly updated, a short book on relativity that he promised in the very title to be *gemeinverständlich* (commonly understandable).[28] It is, however, overly condensed for the most nonscientific beginners. Later, Einstein collaborated with Leopold Infeld in a second attempt to reach out to the population at large by means of a book-length treatment of modern physics. As the preface acknowledged, the authors no longer attempted "a systematic course in elementary facts and theories." Rather, they aimed at a historical account of how the ideas of relativity and quanta entered science, "to give some idea of the eternal struggle of the inventive human mind for a fuller understanding of the laws governing physical phenomena."[29] In fact, there is to this day no generally agreed upon source, the *reading* of which by itself will bring a large fraction of its nonscientific audiences to a sound enough understanding of these ideas, even for those who truly want to attain it and are willing to pay close attention. I believe it is a fact of great consequence that it takes a much larger effort, and one starting earlier than most people assume. To make matters worse, so little has been found out about how scientific literacy is achieved or resisted that not much blame can be spun off on the would-be students, young or old.

Going further along the spectrum in the same direction, we encounter outright hostility and opposition to Einstein's work, either on scientific or on ideological grounds. Almost all scientists, even those initially quite reluctant, became eventually at least reconciled to Einstein's ideas, save (to this day) for his famous refusal to regard the statistical interpretation as fundamental. The opposition to Einstein's work on grounds other than scientific has had a longer history. Thus, a number of studies now exist that show the lengths to which various totalitarian groups, for various reasons, felt compelled to go in their attacks.

Turning now to the other, more "positive" half of the spectrum, we see there the gradual acceptance and elaboration of Einstein's work within the corpus of physical science; its penetration into technology (largely unmarked) and into the more thoughtful philosophies of science; Einstein's effect through his personal intervention, causing some historic redirections of research; and its passage into the scientific world-picture of our time, as it tries to achieve a unification that eluded Einstein. And beyond that, at the end of the spectrum where the number of cases is small but the color deep and vibrant, we perceive the examples of creative transformation beyond science. Those are the works of the few who found that scientific ideas, or rather *metaphors* embodying such ideas, released in them a fruitful response with an authenticity of its own, far removed from textbook physics.

This last is the oldest and surely still the most puzzling interplay between science and the nonscientific aspects of culture. Evidently, the mediation occurs through a sharing of an analogy or metaphor—irresistible despite the dangers inherent in the obvious differences or discontinuities between respective disciplines. We know that such a process exists, because any major work of science itself, in its nascent phase, is connected not only logically, but also analogically both with the historic past in that science and with its supporting data. The scientist's proposal may fit the facts of nature as a glove fits a hand, but the glove does not uniquely imply the existence of the hand, nor the hand that of the glove.

Einstein spoke insistently over the decades about the need to recognize such a discontinuity, one that in his early scientific papers asserted itself first in his audacious method of postulation. In essay after essay, he tried to make the same point, even though it had little effect on the then reigning positivism. Typical are the passages in his Herbert Spencer Lecture of 1933,[30] which we shall examine in some detail in the next chapter. The principles of a theory cannot be "deduced from experience" by "abstraction," that is, by logically complete claims of argument. The creative imagination has to intervene.

If this holds for the creative act in the individual's pursuit of science itself, we should hardly be surprised to find the claim to be extended to more humanistic enterprises. The test, in both instances, is of course whether experience bears out the suitability of the imaginative act. The existence of both splendid scientific theories and splendid products of the humanistic imagination shows that despite all their other differences, they share the ability to build on concepts that, as Einstein put it, are initially of a "purely fictional character." And even the respective fundamentals, despite all their differences, can share a common origin. That is to say, at a given time the cultural pool contains a variety of themata and metaphors, and some of these have a plasticity and applicability far beyond their initial provenance. The innovator, whether a scientist or not, necessarily dips into this pool for his fundamental notions and in turn may occasionally deposit into it new or modified themata and metaphors of general power. Examples of such science-shaped metaphors, each of these by no means a "fact" of the external world yet revealing immense explanatory energy, are easy to find: Newton's "innate force of matter (*vis insita*)" and the Newtonian clockwork universe; Faraday's space-filling electric and magnetic lines of force; Niels Bohr's examples of complementarity in physics and in daily life; Mendeleev's neat tableau setting for the families of elements and Rutherford's lengthy parent-daughter-granddaughter chains of decaying atoms; Minkowski's space-time "World," of which our perceptible space and time are like shadows playing on the wall of Plato's cave; and of course the imaginative scenes Einstein referred to—the traveler along the light beam, the calm experimenter in the freely falling elevator, the dark, dice-playing God, the closed but unbounded cosmos, the Holy Grail of complete unification of all forces of nature. So it continues in science.

The allegorical use of such conceptions may, as we have noted, help to shape works of authenticity outside the sciences. And the process works both ways. Thus Niels Bohr acknowledged that his reading in Kierkegaard and William James helped him to the imaginative leap embodied in his physics; Einstein stressed the influence on his early scientific thinking of the philosophical tracts of that period; and Heisenberg noted the stimulus of Plato's *Timeaus*, read in his school years. No matter if such "extraneous" elements are eventually suppressed or forgotten—or even have to be overcome—at an early point they can encourage the mind's struggle along previously unforeseen ways.

We conclude, then, that in pursuing the evident and documentable cases of "impact" of one person or field—such as that of Einstein and his science—on others, we have been led to a more mysterious fact, namely, the mutual adaptation and resonance of the innovative mind with portions of the total set of themata and metaphors current at a given time. The philosopher José Ortega y Gasset was one of those who struggled with this idea. In 1921–1922, evidently caught up by the rise of the new physics, he began an essay on "The Historical Significance of the Theory of Einstein."[31] There he remarked quite correctly that the most relevant issue was not that the triumph of the theory will influence the spirit of mankind by imposing on it the adoption of a definite route. . . . What is really interesting is the inverse proposition: the spirit of man has set out, of its own accord, upon a definite route, and it has therefore been possible for the theory of relativity to be born and to triumph. The more subtle and technical ideas are, the more remote they seem from the ordinary preoccupations of men, the more authentically they denote the profound variations produced in the historical mind of humanity.[32]

To this day, attempts to go much beyond that point have not been successful. The tantalizing task of finding the detailed, hidden causal links between major works and the spirit of the time at its fruition remains for future scholars.

Chapter 7

E I N S T E I N   A N D   T H E   G O A L

O F   S C I E N C E

THE NOTION OF scientific progress, as noted in Chapters 1 and 2, is today being challenged by various factions outside the laboratories. As to the scientists themselves, they rarely take notice of these currents; and if pressed hard, they would tend to associate themselves with the assertion of George Sarton, the first of the modern historians of science:

> *Definition*: Science is systematized positive knowledge or what has been taken as such at different ages and in different places.
>
> *Theorem*: The acquisition and systematization of positive knowledge are the only human activities which are truly cumulative and progressive.
>
> *Corollary*: The history of science is the only history which can illustrate the progress of mankind. In fact, progress has no definite and unquestionable meaning in fields other than the fields of science.[1]

Apart from tending to sympathize with this sentiment, scientists are aware that scientific activity may be divided roughly into two kinds: that which is directed toward *analysis* and *accretion*, and that which is concerned with *synthesis*. For work of the former kind—by far the larger fraction—the French scientist, historian, and philosopher Pierre Duhem's description applies:

> Physics makes progress because experiment constantly causes new
> disagreements to break out between laws and facts, and because
> physicists constantly touch up and modify laws in order that they
> may more faithfully represent the facts.[2]

The second type of scientific activity, synthesis, is more transfor-
mative. Here progress is equivalent to an increase in inclusiveness (a
wider range of phenomena is accounted for by the new theory) or an
increase in parsimony or restrictiveness (fewer separate fundamental
terms and assumptions are needed). An exemplar, as noted previously, is
the heliocentric system of Copernicus. By virtue of its combination of
inclusiveness and parsimony, Copernicus's system of planets had, as he
explained, the great merit that "not only [do the] phenomena follow
therefrom, but also the order and size of all the planets and spheres and
heaven itself are so linked together that in no portion of it can anything
be shifted without disrupting the remaining parts and the universe as a
whole." In his system, nothing is arbitrary; there is no room for the ad
hoc rearrangement of any orbit, as had been possible before his work.

The Galilean synthesis of celestial and terrestrial physics is another example, as is Newton's *Principia*, particularly in its theory of
lunar motion. Newton's gravitational analysis allowed the prediction,
from fundamentals, of the periods and magnitudes of known "inequal-
ities." His success put forward a vision that laws incorporating such
concepts as force and mass and the use of the proper mathematical and
observational techniques would suffice to discover an overwhelming
unity of physical science (one which Newton privately thought would
eventually encompass the sphere of moral philosophy).

In the twentieth century, the incorporation of Newtonian mechanics into the general theory of relativity was a further example of
scientific progress by synthesis. Similarly, at the end of his *Principles of
Quantum Mechanics*, P. A. M. Dirac confidently declared that since
"quantum mechanics may be defined as the application of equations of
motion to atomic particles," the domain of applicability of that theory
"includes most of low-energy physics and chemistry." These, he implied, were now in principle solved problems.[3]

In almost every scientific research journal, this model of scientific
progress by means of greater inclusiveness and parsimony can be
found. For example, in an article in *Science*, Victor Weisskopf chose six

physical constants—the mass of the proton, the mass of the electron and its electric charge, the velocity of light, Newton's gravitational constant, and the quantum of action of Planck—and a few fundamental laws (for example, de Broglie's relations connecting momentum and energy with wavelength and frequency, and the Pauli principle). Putting these data and laws together, he *derived* from them a host of predictions that correspond to facts of observation: the size and energy of nuclei, the mass and hardness of solids such as rocks, the height of mountains, and the size of our sun and similar stars.[4] All these seemed at first unconnected, as they are spread over an enormous range of magnitudes. Now they are seen to be different consequences of a small number of fundamental posits.

This is indeed Newton's program, the search for omniscience triumphant. And if any one of the six physical constants on Weisskopf's list, now considered a separate entity, could be derived from the other five—for example, if the charge of the electron were found to be deducible from some of the other fundamental constants—that would mark further progress, indeed a great step forward.

A by-product of scientific progress achieved through the continual decrease in the number of fundamental axioms and, simultaneously, through the increase in the range of phenomena covered is an emerging synergism with respect to techniques and methods. Results in one field become useful in another, far distant one, and often in a surprising way. The theoretical tools of condensed-matter physics that handle macroscopic phenomena such as ferromagnetism, superconductivity, and superfluidity are now understood in terms of concepts and theories applicable at the submicroscopic level, where they connect with problems of the structure of stellar bodies, nuclear physics, particle physics, and field theories. Robert Sachs wrote:

> Symmetry principles, and the concepts of gauge theories and spontaneously broken symmetries, turn out to play a common role so that theorists working in one such field may borrow methods from another. Just as all fields have shared in common methods of solution of linear equations, it now appears that new approaches to the solution of non-linear equations, discovered first in connection with hydrodynamic problems, may be of great importance for the understanding of both condensed matter and elementary particles.[5]

THE TWO-DIMENSIONAL MODEL

Instead of attempting to amend the largely intuitive models current among most scientists, or the predominantly ahistorical models among scholars outside science, let us turn to the more useful task of examining a specific case, one which will allow us to obtain a more disciplined model of scientific progress from the practice of a scientist who was knowledgeable about and sensitive to the philosophical underpinnings of modern science. For this purpose there seems to be no better point of departure than the scientist whose work has determined the predominant direction of scientific advance in his field from the early part of the twentieth century and whose approach has been absorbed, consciously or not, by most current practitioners. I mean, of course, Albert Einstein. Einstein wrote extensively on the methods and direction of scientific development, and an appropriate entry for formulating a theory of scientific advance is his essay of 1933, "Zur Methodik der theoretischen Physik."[6] Philipp Frank, Einstein's biographer and colleague, called this the "finest formulation of his views on the nature of a physical theory."[7]

Einstein begins with an important warning, but one which, by itself, would seem to deny the possibility that scientific progress could be achieved on any model: He notes first of all to pay "special attention to the relation between the content of a theory," on the one hand, and "the totality of empirical facts," on the other. These constitute the two "components of our knowledge," the "rational" and the "empirical"; these two components are "inseparable"; but they stand also, Einstein warns, in "eternal antithesis." To support this conception, Einstein refers to a dichotomy built into Western science. The Greek philosopher-scientists provided a lasting reason for having confidence in the achievement of the human intellect by introducing into Western thought the "miracle of the logical system." Logic, as in Euclid's geometry, "proceeds from step to step with such precision that every single one of its propositions was absolutely indubitable." But "propositions arrived at by purely logical means are completely empty as regards reality"; "through purely logical thinking we can attain no knowledge whatsoever of the empirical world." Einstein tells us that it required the seventeenth-century scientists to show that scientific knowledge "starts from experience and ends with it."

Up to this point, therefore, we are left with a thoroughly dualistic method for doing science. On the one hand, Einstein says, "the structure of the system is the work of reason"; on the other hand, "the empirical contents and their mutual relations must find their representation in the conclusions of the theory." Indeed, virtually all of Einstein's commentators have followed him in stressing this dualism—and have left it at that. It is a common view of science.

I consider it a two-dimensional view. It can be defended up to a point and may be summarized as follows. Scientific discourse characteristically deals with two types of meaningful statements, namely, propositions concerning empirical matters that ultimately boil down to meter readings and other public phenomena, and propositions concerning logic and mathematics that ultimately boil down to tautologies. The first of these, the propositions concerning empirical matters of fact, can in principle be rendered in protocol sentences in ordinary language that command the general assent of the scientific community. I call these the *phenomenic* propositions. The second type of propositions, meaningful insofar as they are consistent within the system of accepted axioms, I call *analytic* propositions. As a mnemonic device, and also to do justice to Einstein's warning about the "eternally antithetical" nature of these propositions, one may imagine them as arranged along one or the other of two orthogonal (say, x and y) axes which define the two dimensions of a plane within which scientific discourse usually takes place. A scientific statement, in this view, is therefore analogous to an element of area in the plane, and the projections of the statement onto the two axes are the aspects of the statement that can be rendered, respectively, as a protocol of observation (for example, "the needle swings to the left") and as a protocol of calculation (for example, "use vector calculus, not scalars").

Now, it has been the claim of most modern philosophies of science that trace their roots to empiricism or positivism that any scientific statement has "meaning" only insofar as it can be shown to have phenomenic and/or analytic components in this plane. And, indeed, in the past, this procrustean criterion has amputated from science notions such as innate properties, occult principles, and all kinds of tantalizing questions which cannot be expressed in terms of those two dimensions and for which the consensual mechanism consequently could not produce sufficiently satisfying answers. A good argument can be made that the silent but general agreement to keep the discourse consciously in the phenomenic-analytic plane, where statements and routines can be shared, is the main reason science has been able to grow so rapidly in modern times. While the degree of consensus at the developing edge of science is usually far less than most model makers of science realize, consensus about an area rises rapidly after the edge has moved on. Hence, the two-dimensional model is widely used to characterize what is "truly scientific" when writing for pedagogic purposes.

LIMITATIONS OF THE TWO-DIMENSIONAL MODEL

The two-dimensional view has its costs. It overlooks or denies the existence of other active mechanisms at work in the day-to-day experience of those engaged in the pursuit of science; and it is of little help in handling questions every historian of science has to face consciously, even if the working scientist, happily, does not. To illustrate, let me mention two such problems. Both have to do with the direction of scientific advance, and both will seem more amenable to solution once the dualistic view is modified.

First, if sound scientific discourse is directed entirely by the dictates of logic and of empirical findings, why is science not one great totalitarian engine, taking everyone relentlessly in the same way to the same inevitable goal? The laws of reason, the perception or detection of phenomena, and human skills to deal with both are presumably distributed equally over the globe; and yet the story of, say, the reception of Einstein's theories is strikingly different in Germany and England, in France and the United States. On the level of *personal* choice of a research topic, why were some of Einstein's contemporaries so fatally attracted to ether-drift experiments; whereas he himself, as he put it to his friend W. J. de Haas, thought such experiments as silly and doomed to failure as trying to study dreams in order to prove the existence of ghosts? As to skills for navigating in the two-dimensional plane, Einstein and Bohr were rather well matched, as were Schrödinger and Heisenberg. And yet there were fundamental antagonisms in terms of programs, tastes, and beliefs, with occasional passionate outbursts among scientific opponents.

Or, again, how to understand the great variety of different personal styles of scientists, all engaged in what they agree to be the "same" problem? If science *were* two-dimensional, the work of scientists in a given field might sooner or later be governed by a rigid, uniformly accepted exemplar. The documented existence of pluralism at all times points to a fatal flaw in the two-dimensional model—but also to its cure.

A second question that escapes the simple model, and to which I have referred in Chapters 4 and 5, is this: why are many scientists, particularly in the nascent phase of their work, willing to hold firmly, and sometimes at great risk, to what can only be called a suspension of disbelief about the possibility of disconfirmation? Moreover, why do they sometimes do so at the early stages of the search without having any empirical evidence on their side, or even in the face of contradictory evidence?

Among countless examples of this sort, Max Planck, responsible for the idea of the quantum but one of the most outspoken opponents of its corpuscular implications, cried out as late as 1927: "Must we really ascribe to the light quanta a physical reality?"—this four years after the publication of Arthur H. Compton's findings providing conclusive evidence. On the other hand, when it came to explaining the electron in terms of what Planck called "vibrations of a standing wave in a continuous medium," along the lines proposed by de Broglie and Schrödinger, Planck gladly accepted the idea and added that "these principles have already [been] established on a solid foundation"—this before Planck had heard of any experimental evidence for the wave aspect of matter, along the lines provided by Davisson and Germer.[8]

Einstein was even more daring. To select just one among many illustrations, in 1916, when he wrote his book *Über die spezielle und die allgemeine Relativätstheorie*, he had to acknowledge that his general theory of relativity so far had only one observable consequence, the precession of the orbit of Mercury; whereas the predicted bending of light and the red shift of spectral lines owing to the gravitational potential were too small to then be observed. Nevertheless, Einstein readily proclaimed, in a daring sentence with which he ended his book in its first fifteen printings, from 1917 through 1919: "I do not doubt at all that these consequences of the theory will also find their confirmation." It is an example of the suspension of disbelief, an important mechanism in the practice of experimental and theoretical scientists.[9]

THE ROLE OF PRESUPPOSITIONS

What, then, must one conclude from Planck's predisposition for the continuum and against discreteness, from Einstein's predisposition for a theory that encompasses a wide rather than a narrow range of phenomena and so allows him to risk his reputation on a daring prediction, and from many other examples of suspension of disbelief in the face of missing tests and even (in the case of R. A. Millikan's oil-drop experiment) contrary data? Such cases serve to indicate that some *third mechanism* is present in determining the choices scientists make in the nascent phase of their work, in addition to the phenomenic and analytical mechanisms. Indeed, we can find it in Einstein's lecture on the method of theoretical physics: The two-dimensional model initially prominent in it gives way, on closer examination, to a more sophisticated and appropriate one. In addition to the two inseparable but antithetical components, there is indeed a third one, directed away from the plane bounded by the empirical and logical dimensions of the theory.

Einstein launches his argument by reminding his audience, as he often did, that the principles of a theory cannot be "deduced from experience" by "abstraction"—that is to say, by logical means. "In the logical sense [the fundamental concepts and postulates of physics are] free inventions of the human mind," in themselves "completely empty as regards reality," and in that sense different from the unalterable Kantian categories. He repeats more than once that the "concepts with which scientific theories are built are necessarily the products of the human imagination, hence are initially 'purely fictitious' in character."[10] Or as he memorably put it soon afterward: the relationship between sense experience and concept "is analogous not to that of soup to beef [where the broth is simply a direct extract of the meat], but rather to that of check number to overcoat."[11]

This arbitrariness of reference, Einstein explains, "is perfectly evident from the fact that one can point to two essentially different foundations"—the general theory of relativity and Newtonian physics—"both of which correspond with experience to a large extent," namely, with much of mechanics. Moreover, the elementary experiences do not provide a logical bridge to the basic concepts and postulates of either relativity or Newtonian mechanics. Rather, "the axiomatic basis of theoretical physics . . . must be freely invented." (To be sure, eventually experience will decide whether the invention was a useful and appropriate one.)

With this declaration, Einstein has, of course, exposed the emptiness of attempts to impose external standards of correct thinking on the practice of scientists, or to condemn as "irrational" scientific work that fails to meet such criteria. Thus, he shows that good scientific reasoning follows the precepts of neither the Dionysians nor the Apollonians. Einstein is, however, quite aware that his insight leads immediately to a basic problem, and he spells it out: How "can we ever hope to find the right way? Nay, more, has this right way an existence outside our illusions? Can we hope to be guided safely by experience at all when there exist theories such as classical mechanics, which do justice to experience to a large extent, but without grasping the matter in a fundamental way?"

We have now left the earlier, confident portion of Einstein's lecture far behind. But at this very point, Einstein issues a clarion call: "I answer with full confidence that there is, in my opinion, a right way, and that we are capable of finding it." Here, Einstein goes suddenly beyond his earlier categories of empirical and logical efficacy, and offers us a set of selection rules with which, as with a good map and compass, that "right way" may be found. Here, there, everywhere, guiding concepts emerge in his essay and beckon from above the previously defined plane to point us on the right path.

The first directing principle Einstein mentions is his personal belief in the efficacy of *formal structures*. The "creative principle resides in mathematics"—not, for example, in mechanical models. Next there unfolds a veritable hymn to the guiding concept of *simplicity*. Einstein calls it "the principle of searching for the mathematically simplest concepts and their connections," and he cheers us on our way with many examples of how effective it has already proven to be:

> If I assume a Riemannian metric [in the four-dimensional continuum] and ask what are the *simplest* laws which such a metric can satisfy, I arrive at the relativistic theory of gravitation in empty space. If in that moment I assume a vector field or antisymmetrical tensor field which can be derived from it, and ask what are the simplest laws

which such a field can satisfy, I arrive at Maxwell's equations for empty space.

And so on, collecting victories everywhere under the banner of simplicity.

Later in the lecture, we find two other guiding notions in tight embrace: the concept of *parsimony* or *economy*, and that of *unification*. As science progresses, Einstein tells us, "the logical edifice" is more and more "unified," the "smaller [is] the number of logically independent conceptual elements which are necessary to support the whole structure." Higher up on that same page, we encounter nothing less than what he calls "the noblest aim of all theory," which is "to make these irreducible elements as simple and as few in number as is possible, without having to renounce the adequate representation of any empirical content."

Yet another guiding concept to which Einstein gladly confesses is the *continuum*, the field. From 1905 on, when the introduction into physics of discontinuity in the form of light quanta forced itself on Einstein as a "heuristic" and therefore not fundamental point of view, he clung to the hope and program to keep the continuum as a fundamental conception, and he defended it with enthusiasm in his correspondence. It was part of what he called his "Maxwellian program" to fashion a unified field theory. Atomistic discreteness and all it entails was not the solution but rather the problem. So here he again considers the conception of "the atomic structure of matter and energy" to be "the great stumbling block for a unified field theory."

One cannot, he thought, settle for a basic duality in nature, giving equal status both to the field and to its antithesis, discrete matter. To be sure, neither logic nor experience forbade it. Yet for him it was unthinkable. As he wrote to his old friend, Michele Besso, "I consider it quite possible that physics might not, finally, be founded on the concept of field—that is to say, on continuous elements. But then out of my whole castle in the air—including the theory of gravitation, but also most of current physics—there would remain *nothing*."[12]

We have by no means come to the end of the list of presuppositions that guided Einstein. But it is worth pausing to note how plainly he seemed to have been aware of their operation in his own scientific work. In this too he was rare. Isaiah Berlin, in his book *Concepts and Categories*, remarked: "The first step to the understanding of men is the bringing to consciousness of the model or models that dominate and penetrate their thought and action. Like all attempts to make men aware of the categories in which they think, it is a difficult and sometimes painful activity, likely to produce deeply disquieting results."[13] This is generally true; but it was not so difficult for Einstein, for at least two reasons. It was he, after all, who first realized the "arbitrary character" of what had for so long been accepted as "the axiom of the absolute character of time, viz., of simultaneity [which] unrecognizedly was anchored in the unconscious," as he put it in his "Autobiographical Notes." "Clearly to recognize this axiom and its arbitrary character really implies already the solution of the problem."[14] Having to give up an explicitly or implicitly held presupposition has indeed often had the characteristic of the great sacrificial act of modern science. We find in the writings of Kepler, Planck, Bohr, and Heisenberg that such an act climaxes a period that in retrospect is characterized by the word "despair."

Having recognized and overcome the negative, or enslaving, role of presuppositions, Einstein also saw their positive, emancipating potential. In one of his earliest essays on epistemology, he wrote:

> A quick look at the actual development teaches us that the great steps forward in scientific knowledge originated only to a small degree in this [inductive] manner. For if the researcher went about his work without any preconceived opinion, how should he be able at all to select out those facts from the immense abundance of the most complex experience, and just those which are simple enough to permit lawful connections to become evident?[15]

Much later, in his "Reply to Criticisms," appended to his "Autobiographical Notes," he reverted to the "eternal antithesis" by way of acknowledging that the distinction between "sense impressions," on the one hand, and "mere ideas," on the other, is a basic conceptual tool for which he can adduce no convincing evidence. Yet he needs this distinction. His solution is simply to announce: "We regard the distinction as a category which we use in order that we might the better find our way in the world of immediate sensation." As with other conceptual distinctions for which "there is also no logical-philosophical justification," one has to accept it as "the presupposition of every kind of physical thinking," mindful that "the only justification lies in its usefulness. We are here concerned with 'categories' or schemes of thought, the selection of which is, in principle, entirely open to us and whose qualification can only be judged by the degree to which its use contributes to making the totality of the contents of consciousness 'intelligible.' " Finally, he curtly dismisses an implied attack on these "categories" or "free conventions" with the remark: "Thinking without the positing of categories and concepts in general would be as impossible as is breathing in a vacuum."[16]

FURTHER DETAILS ON THE USE OF THEMATA

Einstein's remarkable self-consciousness concerning his fundamental presuppositions throughout his scientific and epistemological writings allows one to assemble a list of about ten chief presuppositions underlying his theory of construction throughout his long scientific career: primacy of formal (rather than materialistic or mechanistic) explanation; unity or unification; cosmological scale in the applicability of laws; logical parsimony and necessity; symmetry (for as long as possible); simplicity; causality (in essentially the Newtonian sense); completeness and exhaustiveness; continuum; and, of course, constancy and invariance.

These ideals, to which Einstein was fiercely devoted, explain why he would continue his work even when tests against experience were difficult or unavailable, or, conversely, why he refused to accept theories supported by the phenomena but, as in the case of Bohr's quantum mechanics, based on presuppositions opposite to his own. As has been touched on in earlier chapters, much of the same can be said of most major scientists, from Johannes Kepler and Galileo to our contemporaries. Each has his own, sometimes idiosyncratic, "map" of fundamental guiding notions that may be considered in principle separate, like the band structure of chromosomes in the nucleus of a cell.

With this finding, we must now reexamine our mnemonic device of the two-dimensional plane. I remove its insufficiency by defining a third axis, rising perpendicularly out of it. This is the dimension orthogonal to and not resolvable onto the phenomenic or analytic axes. Along it are located those fundamental presuppositions, often stable, many widely shared, that show up in the motivation of the scientist's actual work, as well as in the end product for which he or she strives, and in the acceptance or rejection of scientific insights. A scientist's choices among the presuppositions, insofar as they are consciously made, are judgmental (rather than, as in the phenomenic-analytic plane, capable in principle of algorithmic decidability). Since these fundamental presuppositions are not directly derivable either from observation or from analytic ratiocination, they require a term of their own. I have called them *themata* (singular *thema*, from the Greek θέμα: meaning "that which is laid down, proposition, primary word").

On this view—and again purely as a mnemonic device—a scientific statement is no longer, as it were, an element of area on the two-dimensional plane but a volume-element, an entity in three-dimensional space, with components along each of the three orthogonal axes (x, y, and z, respectively the phenomenic, analytic, and thematic axes). The projection of the entity onto the two-dimensional, x-y plane continues to have the useful roles I stressed earlier; but for our analysis it is also necessary to consider the line element projected onto the third, or z axis, the dimension on which one may imagine the range of themata to be entered. The statements of differing scientists are therefore like two volume-elements that do not completely overlap and so have some differences in their projections.

A scientist's thematic attachment may favor one or the other of two (or sometimes three) antithetical themata (i.e., θ or $\bar{\theta}$), and these are often traceable to very early speculations, such as those concerning constancy versus change in the traditions of Parmenides and Heraclitus. The historian of science trying to understand a specific case or event should therefore be alert to possible thematic choices made by the scientist: choices between experience and symbol formation, complexity and simplicity, reductionism and holism, discontinuity and continuum, hierarchical levels and unity, evolution/devolution/steady state, mechanistic/materialistic/mathematical models, causality/probabilism, analysis/synthesis, and so on.

While scientists are generally not, and need not be, conscious of the themata they use, the historian of science can chart the growth of a given thema in the work of an individual scientist over time and show its power upon the scientist's imagination. Thematic analysis, then, is in the first instance the identification of the particular map of the various themata which, like fingerprints, can characterize an individual scientist, or a part of the scientific community, at a given time.

Most themata are not only old but long-lived, and show up most strikingly during a conflict between individuals or groups that base their work on opposing presuppositions. I have been impressed by the rather small number of thematic couples, or triads. A relative few have sufficed us throughout the history of the physical sciences. Thematic analysis of the same sort has begun to be brought to bear also on significant cases in other fields.[17]

With this conceptual tool we can return to some of the puzzles we mentioned earlier. Where does the conceptual and even emotional support come from which, for better or worse, stabilizes the individual scientist's risky speculation and confident suspensions of disbelief during the nascent phase? The result of case studies is that choices and decisions of this sort are often made on the basis of loyal dedication to thematic presuppositions. Or again, if, as Einstein claimed, the principles are indeed initially free inventions of the human mind, should those not be an infinite set of possible axiom systems to which one could leap or cleave? Virtually every one of these would ordinarily be useless for constructing theories. How then could there be any hope of success, except by chance? The answer must be that the license implied in the leap to an axiom system of a theory by the freely inventing mind is the freedom to make such a leap, but not the freedom to make *any leap whatever*. The freedom is narrowly circumscribed by a scientist's particular set of themata that constrains and shapes the style, direction, and rate of advance on novel ground.

And insofar as the individual sets of themata overlap, the progress of the scientific community as a group is similarly constrained or directed. Otherwise, the inherently anarchic connotations of "freedom" could indeed disperse the total effort. Mendeleev wrote: "Since the scientific world view changes drastically not only from one period to another but also from one person to another, it is an expression of creativity. . . . Each scientist endeavors to translate the world view of the school he belongs to into an indisputable principle of science." In practice, however, there is more coherence than this statement implies, and we shall presently look more closely at the mechanism responsible for it.

THE "NEED TO GENERALIZE"

Of all the problems that invite investigation with these tools, the most fruitful example will be a return visit to that mysterious place, early in the 1933 essay, where Einstein speaks of the need to pay "special attention to the relations between the content of the theory and the totality of empirical fact." The *totality* of empirical fact! It is a phrase that recurs in his writings and indicates the sweep of his ambition. But it does even more: It lays bare the most daring of all the themata of science and points to the holistic drive behind "scientific progress."

Einstein explicitly and frankly hoped for a theory that would ultimately be utterly comprehensive and completely unified. This vision drove him on from the special to the general theory, and then to the unified field theory. The search for one grand, architectonic structure itself was of course not Einstein's invention. On the contrary, it is an ancient dream. At its worst, it has sometimes produced authoritarian visions that are as empty in science as their equivalent is dangerous in politics. At its best, it has propelled the drive toward the various grand syntheses that rise above the more monotonous landscape of analytic science. This has certainly been the case in recent decades in the physical sciences. Today's defenders of the promise as applied to particle physics, who in the titles of their publications use the term "Grand Unification," are in a real sense the hopeful children of those earliest synthesis-seekers of physical phenomena, the Ionian philosophers.

To be sure, as Isaiah Berlin warned in *Concepts and Categories*, the quest for greater synthesis, successful so far, from Oersted to Maxwell and from Einstein to our day, may be a trap. Berlin has christened it the "Ionian Fallacy," defined as the search, from Aristotle to Bertrand Russell to our own day, for the ultimate constituents of the world. Superficially, the seekers of a unified physics, particularly in their monistic exhortations, may seem to have risked falling into that trap— from Copernicus, who confessed that the chief point of his work was to perceive nothing less than "the form of the world and the certain commensurability of its parts," to Max Planck, who exclaimed in 1915 that "physical research cannot rest so long as mechanics and electrodynamics have not been welded together with thermodynamics and heat radiation,"[18] to today's theorists who seem to follow the founding father of science among the ancient Greeks, Thales himself, in their insistence that one entity will explain all.

A chief point in my view of science is that by the actual variety of the individual scientists' thematic commitments, the community of scientists is in practice rescued from the trap that devotion to a single thema might eventually lead to. The multiplicity of their views gives them as a group the flexibility that an authoritarian research program built on a single thema lacks. (There is an analogy here with the benefits of maintaining biodiversity among crop plants.)

Without doubt, something like an Ionian Enchantment, the commitment to the theme of grand unification, was upon Einstein. Once alerted, we can find it in his work from the very beginning, in his first published paper (1901), where he tried to understand the contrary-appearing forces of capillarity and gravitation. We noted in Chapter 5 his revealing confession at that early stage that for him it was "a magnificent feeling to recognize the unity [*Einheitlichkeit*] of a complex of phenomena that to direct observation appear to be quite separate things"—in this case the physics of micro- and macro-regions. In each of his next papers we find some of the same drive, which he later called "my need to generalize." He examined whether the laws of mechanics provide a sufficient foundation for the general theory of heat and whether the fluctuation phenomena that turn up in statistical mechanics also explain the basic behavior of light beams and their interference, the Brownian motion of microscopic particles in fluids, and even the fluctuation of electric charges in conductors. And in his deepest work of those early years, in special relativity theory, the most powerful propellant was Einstein's drive toward unification. His clear motivation was to find a more general point of view that would subsume the seemingly limited and contrary problems and methods of mechanics and of electrodynamics. In the process he showed that electric and magnetic fields are aspects of one commonality viewed from different reference frames: that space and time are not separate;

that energy and mass are fused in one conservation law; and, soon after, that reference systems with gravitation and with acceleration are equivalent. Again and again, previously separate notions were shown to be connected.

Following the same program obstinately to the end of his life, he tried to bring together, as he had put it once, "the gravitational field and the electromagnetic field into a unified edifice," leaving "the whole physics" as a "closed system of thought."[19] In his longing for a unified world picture—a structure that would yield deductively the "totality" of empirical facts—one cannot help hearing an echo of Goethe's Faust, who exclaimed that he longed "to detect the inmost force that binds the world and guides its course." For that matter, one hears Newton himself, who wanted to build a unifying structure so tight that the most minute details would not escape it.

THE UNIFIED *WELTBILD* AS "SUPREME TASK"

In its modern form, the Ionian Enchantment, expressing itself in the search for a unifying world picture, is usually traced to Alexander von Humboldt and Schleiermacher, Fichte and Schelling. The influence of the "Nature philosophers" on physicists such as Hans Christian Oersted—who was led by their ideas directly to the first experimental demonstration of the unification of electricity and magnetism—has been amply chronicled. At the end of the nineteenth century, in the Germany of Einstein's youth, the pursuit of a unified world picture as the scientist's highest task had become almost a cult activity. Looking on from the other side of the channel, J. T. Merz exclaimed in 1904 that the lives of the continental thinkers are

> devoted to the realization of some great ideal. . . . The English man of science would reply that it is unsafe to trust exclusively to the guidance of a pure idea, that the ideality of German research has frequently been identical with unreality, that in no country has so much time and power been frittered away in following phantoms, and in systematizing empty notions, as in the Land of the Idea.[20]

Einstein himself could not easily have escaped notice of these drives toward unification even as a young person. For example, we know that as a boy he was given Ludwig Büchner's widely popular book *Kraft und Stoff* [*Energy and Matter*], a work Einstein recollected having read with great interest. The little volume does talk about energy and matter; but chiefly it is a late-Enlightenment polemic. Büchner comes out explicitly and enthusiastically in favor of an empirical, almost Lucretian scientific materialism, which Büchner calls a "materialistic world view." Through this worldview, he declares, one can attain "the unity of energy and matter, and thereby banish forever the old dualism."[21]

But the books which Einstein himself credited as having been the most influential on him in his youth were Ernst Mach's *Theory of Heat* and *Science of Mechanics*. That author was motivated by the same Enlightenment ideals and employed the same language. In the *Science of Mechanics*, Mach exclaims: "Science cannot settle for a ready-made world view. It must work toward a future one . . . that will not come to us as a gift. We must earn it! [At the end there beckons] the idea of a unified world view, the only one consistent with the economy of a healthy spirit."[22]

Indeed, in the early years of the twentieth century, German scientists were thrashing about in a veritable flood of publications that called for the unification or reformation of the "world picture" in the very title of their books or essays. Planck and Mach carried on a bitter battle, publishing essays directly in the *Physikalische Zeitschrift* with titles such as "The Unity of the Physical World Picture." Friedrich Adler, one of Einstein's close friends, wrote a book with the same title, attacking Planck. Max von Laue countered with an essay he called "The Physical World Picture." The applied scientist Aurel Stodola, Einstein's admired older colleague in Zurich, corresponded at length with Einstein on a book which finally appeared under the title *The World View of an Engineer*. Similarly titled works were published by other collaborators and friends of Einstein, such as Ludwig Hopf and Philipp Frank.

Perhaps the most revealing document of this sort was the manifesto published in 1912 in the *Physikalische Zeitschrift* on behalf of the new *Gesellschaft für positivistische Philosophie*; it had been composed in 1911 at the height of the *Weltbild* battle between Mach and Planck. Its declared aim was nothing less than "to develop a comprehensive *Weltanschauung*," and thereby "to advance toward a noncontradictory, total conception [*Gesamtauffassung*]." The document was signed by, among others, Ernst Mach, Josef Petzoldt, David Hilbert, Felix Klein, George Helm, Albert Einstein (only just becoming more widely known at the time), and that embattled builder of another worldview, Sigmund Freud.[23]

It was perhaps the first time that Einstein signed a manifesto of any sort. That it was not a casual act is clear from his subsequent, persistent return to the same theme. His most telling essay was delivered in late 1918, possibly triggered in part by the publication of Oswald Spengler's *Decline of the West*, that polemic against what Spengler called "the scientific world picture of the West." Einstein took the occasion of a presentation he made in honor of Planck (in *Motiv des Forschens*) to lay out in detail the method of constructing a valid world picture. He insisted that it was not only possible to form for oneself "a simplified world picture that permits an overview [*übersichtliches Bild der Welt*]," but that it was what he called the scientist's "supreme task to do so." Specifically, the worldview of the theoretical physicist "deserves its proud name *Weltbild* because the general laws upon which the conceptual structure of theoretical physics is based can assert the claim that they are valid for any natural event whatsoever. . . . The supreme task of the physicist is therefore to seek those most universal elementary laws from which, by pure deduction, the *Weltbild* may be achieved."[24]

There is of course no doubt that Einstein's work during those years constituted great progress toward this self-appointed task. In the developing relativistic *Weltbild*, a huge portion of the world of events and processes was being subsumed in a four-dimensional structure which Hermann Minkowski in "*Raum und Zeit*" (1908) named simply "*die Welt*"—the world conceived as a Parmenidean crystal made of world lines, in which changes, such as motions, are largely suspended. In this world, the main themata are those of constancy and invariance, determinism, necessity, and completeness.

LEAVING OUT NOT A SINGLE EVENT

Typically, it was Einstein himself who knew best, and recorded frequently, the limitations of his work. Even as special relativity began to make converts, he announced that the solution was quite incomplete because it applied only to inertial systems and left out entirely the great puzzle of gravitation. Later he worked on removing the obstinate dualities, explaining for example that "measuring rods and clocks would have to be represented as solutions of the basic equation . . . not, as it were, as theoretical self-sufficient entities." This he called a "sin" which "one must not legalize." The removal of the sin was part of the hoped-for perfection of the total program, the achievement of a unified field theory in which "the particles themselves would *everywhere* be describable as singularity-free solutions of the complete field-equations. Only then would the general theory of relativity be a *complete* theory."[25] Therefore the work of finding those most general elementary laws from which by pure deduction a single, consistent, and complete *Weltbild* can be won had to continue.

There has always been a notable polarity in Einstein's thought with respect to the completeness of the world picture he was seeking. On the one hand, he insisted from beginning to end that no single event, individually considered, must be allowed to escape from the final grand net. We have already noted that in the Oxford lecture of 1933 he was concerned with encompassing the "totality of experience," and he declared the supreme goal of theory to be "the adequate representation of any content of experience."[26] He even goes beyond that: Toward the end of his lecture he reiterates his old opposition to the Bohr-Born-Heisenberg view of quantum physics and declares, "I still believe in the possibility of a model of reality, that is to say a theory, which shall represent the events themselves [*die Dinge selbst*] and not merely the probability of their occurrence." Writing three years later he insists even more bluntly:

> But now, I ask, does any physicist whosoever really believe that we shall never be able to attain insight into these significant changes of single systems, their structure, and the causal connections, despite the fact that these individual events have been brought into such close proximity of experience, thanks to the marvelous inventions of

the Wilson-Chamber and the Geiger counter? To believe this is, to be sure, logically possible without contradiction; but it is in such lively opposition to my scientific instinct that I cannot forgo the search for a more complete mode of conception.[27]

Yet even while Einstein was anxious not to let a single event escape from the final *Weltbild,* he seems to have been strangely uninterested in nuclear phenomena, that lively branch of physics which began to command great attention precisely in the years Einstein started his own researches. He seems to have thought that these phenomena, in a relatively new and untried field, would not soon lead to the deeper truths. And one can well argue that he was right; not until the 1930s was there a reasonable theory of nuclear structure, and not until after the big accelerators were built were there adequate conceptions and equipment for the hard tests of the theories of nuclear forces.

Einstein's persistent pursuit of a fundamental theory, one so powerful that it would include every datum of experience and yet excluded nuclear phenomena, can be understood only as a consequence of a suspension of disbelief of an extraordinary sort. It is ironic that, as it turned out, even while Einstein was trying to unify the two long-range forces (electromagnetism and gravitation), the nucleus was harboring two additional fundamental forces and, moreover, that after a period of neglect, the modern unification program, two decades after Einstein's death, began to succeed in joining one of the nuclear (relatively short-range) forces with one of the relatively long-range forces (electromagnetism). In this respect, the landscape through which the scientists have been moving appears now to be less symmetrical than Einstein had thought it to be.

For this and similar reasons, few of today's working researchers consciously identify their drive toward "grand unification" with Einstein's. Their attention is attracted by the thematic differences, expressed for example by their willingness to accept a fundamentally probabilistic world. And yet the historian can see the profound continuity. Today, as in Einstein's time and indeed that of his predecessors, the majority of fundamental research physicists hope for the achievement of a logically unified and parsimoniously constructed system of thought that will provide the conceptual comprehension, as complete as humanly possible, of the scientifically accessible sense experiences in their full diversity. This ambition embodies a final aim of scientific work itself, and it has done so since the rise of science in the Western world. Most scientists, working on small fragments of the total structure, are as unself-conscious about their participation in that grand monistic task as they are about, say, their monotheistic assumption, which remains central to their personal belief system without demanding explicit avowal. Indeed, Joseph Needham may well be right that the development in the West of the concept of a unified natural science depended on the preparation of the ground through monotheism, so that one can understand more easily the reason that modern science arose in seventeenth-century Europe rather than, say, in China.

THEMATIC PLURALISM AND THE DIRECTION OF PROGRESS

Diversity in the spectrum of themata held by individual scientists, and overlap among these sets of themata: This formula seems to me to answer the question of why the preoccupation with the eventual achievement of a unified world picture did not lead physics to a totalitarian disaster, as an Ionian Fallacy by itself could well have done. At every step, each of the various world pictures in use is considered a preliminary version, a premonition of the Holy Grail. Moreover, each of these various hopeful but incomplete world pictures that guide scientists at a given time is not a seamless, unresolvable entity. Nor is each completely shared even within a given subgroup. Each member of the group operates with a specific spectrum of separable themata, some of which are also present in portions of the spectrum in rival world pictures. Einstein and Bohr agreed far more than they disagreed, but they did have profound thematic differences. Moreover, most of the themata current at any one time are not new but are adopted from predecessor versions of the *Weltbild*, just as many of them would later be incorporated in subsequent versions of it. Einstein freely called his project a "Maxwellian program" in this sense.[28]

It is also for this reason that Einstein saw himself with characteristic clarity not at all as a revolutionary, as his friends and his enemies so readily did. He took every opportunity to stress his role as a link in an evolutionary chain. Even while he was working on relativity theory in 1905, he called it "a modification" of the theory of space and time.

Later, in the face of being acclaimed as the revolutionary hero of the new science, he insisted, as in his King's College lecture: "We have here no revolutionary act but the natural development of a line that can be traced through centuries." Relativity theory, he held, "provided a sort of completion of the mighty intellectual edifice of Maxwell and Lorentz."[29] Indeed, he shared quite explicitly with Maxwell and Lorentz some fundamental presuppositions, such as the need to describe reality in terms of continua (fields), even though he differed from them completely with respect to others, such as the role of a plenum (ether).

On this model of the role which the thematic component plays in the advancement of science, we can now understand why scientists do not and need not hold substantially the same set of beliefs, either to communicate meaningfully with one another in agreement or disagreement or in order to contribute to cumulative improvement of the state of science. Their beliefs have considerable fine structure; and within that structure there is room both for thematic overlap and agreement, which generally have a stabilizing effect, and for intellectual freedom, which may be expressed as thematic disagreements. Innovations emerging from the balance, even the "far-reaching changes," as Einstein termed the contributions of Maxwell, Faraday, and Hertz, require neither from the individual scientist nor from the scientific community the kind of radical and sudden reorientation implied in such currently fashionable language as revolution, Gestalt switch, discontinuity, incommensurability, conversion, etc. On the contrary, the innovations are consistent with the advancement of science as an evolutionary process, to which Einstein himself explicitly adhered and which emerges also from the actual historical study of his scientific work.

Thus, major scientific advances can generally be understood in terms of a process that involves battles over only a few but by no means all of the recurrent themata. The work of scientists acting individually or in a group, seen synchronically or diachronically, is not constrained to the phenomenic-analytic plane alone; it is, rather, an enterprise whose saving pluralism resides in its many internal degrees of freedom. Therefore, we can understand why scientific progress is often disorderly, but not catastrophic; why there are many errors and delusions, but not one great fallacy; and how mere human beings, confronting the seemingly endless, interlocking puzzles of the universe, can advance at all—as advance they have, if not all the way to the Elysium of Einstein's own hope, the single world conception that grasps the totality of phenomena.

Chapter 8

OF PHYSICS, LOVE,

AND OTHER PASSIONS:

THE LETTERS OF ALBERT

AND MILEVA

POPULAR OPINION HAS it that the scientist is, and should be, *dispassionate*, implying that he or she is somehow less than human. But when historians of science look at the stages of a scientist's work prior to publication, a different picture emerges. Depending on the case, dispassion may be the least of the virtues in evidence. Even the motivations for persisting in the exhausting pursuit of a difficult problem would be inexplicable without the operation of a whole range of emotions that banal opinion would grant to everyone except scientists. And when it comes to studying the private letters of certain scientists (such as Erwin Schrödinger), the evidence of passions ruling both the personal and scientific life can reach volcanic proportions.

One must approach such a case with caution because of the difficulty of trying to put oneself into the frame of mind in which very personal letters were written at some point in the distant past. This problem was put very well by A. S. Byatt in her novel *Possession—A Romance* (1990), a fascinating concoction centering on the discovery of love letters between the fictional Randolph Henry Ash, a mid-nineteenth-century poet and amateur scientist, and the mysterious writer Christabel La Motte. The book proposes that two late-twentieth-century English literature scholars, Roland Mitchell and Maud Bailey, discover that correspondence, with its record of the romance, the mysterious ending of it, the question of whether there was a child and what became of her. Byatt indicates the problem for the contemporary scholar. Maud says:

> "I have been trying to imagine him. Them. They must have been—in an extreme state. . . . We are very knowing. . . . We know we are driven by desire, but we can't see it as they did, can we? We never say the word Love, do we—we know it's a suspect ideological construct—especially Romantic Love—so we have to make a real effort of imagination to know what it felt like to be them, here, believing in these things—love—themselves—that what they did mattered—"

Let's make that effort, for a couple of young people who longed for each other as well as for a life in science.

MILEVA

Mileva Marić was born in 1875. Her father, Miloš, was a Serb, and her mother, Marija Rûzić, came from a Montenegrin family. They were living in Vojvodina, then part of southern Hungary, a few dozen miles northwest of Belgrade, an area that had initially been settled by farmer-warriors to protect Christian Europe from the Turks. Mileva's was a rather well-to-do family, her father being an official in the state bureaucracy. From earliest times, she showed that she would be a serious and dedicated young woman. For example, during her high school years, she sought and received special permission to attend an all-male *Gymnasium* to take two years of physics—a course which otherwise was not available to girls generally. It is reported that she got top grades in physics as well as in mathematics.

It was very difficult for a woman to go on to higher education in most parts of Europe. The first European country to admit women to its universities was France in 1863, followed by the University of Zurich in Switzerland in 1865, with other Swiss universities following a few years later. So if you were French or, like Maria Sklodowska (later Marie Curie), comfortable with the French language, you would naturally move to Paris for university training. If, like Mileva, you were fluent in German, it was more natural to go to Switzerland. Indeed, as John Stachel points out, "Most of the women entering Swiss higher schools during the nineteenth century were non-Swiss, a large proportion of them being of Slavic background."[1]

In 1894, Mileva went to Switzerland, first taking the last two years of secondary school, then doing a summer semester in medical studies at the University of Zurich in 1896, and finally, in the fall of that year, enrolling in the Swiss Federal Polytechnic School (called "Poly" for short, or "ETH," for Eidgenössische Technische Hochschule).

From the testimony of her fellow students and those who knew her soon after, she must have been a formidable young woman, with a fierce dedication to the study of physics. In conversation, she tended to be quiet. In appearance she was darkly handsome. Having been born with a congenital hip dislocation from which her sister also suffered, she walked with a slight limp. One is reminded of studies such as that of sociologist Anne Roe, who found that scientists, far beyond the usual proportion in the general population, had periods of illness or physical handicap in childhood. Some psychologists[2] have related this fact to the prominent incidence of introversion among young persons hoping to become scientists—all such characteristics helping the youth to resist early peer pressure, which generally was away from science as a career. Einstein described another mechanism serving the same end when he wrote in his autobiography that he had passed through an early period of "deep religiosity . . . which . . . found an abrupt ending at the age of twelve, through the reading of popular scientific books." And he added, "It is quite clear to me that the religious paradise of youth, which was thus lost, was a first attempt to free myself from the chains of the 'merely personal,' from an existence which is dominated by wishes, hopes, and primitive feelings."[3]

What were Mileva's career interests? One must remember that the chance of obtaining a university professorship in any science in the 1890s and early 1900s was very small even if one completed the regular academic university preparation. Such positions were few (fifty-four in physics at all university ranks in Austria-Hungary in 1909); openings occurred rarely, and the road to them was long.

Whether for this reason or because her ambitions lay elsewhere, Mileva enrolled in the Poly as a student in Section VIA, which trained future teachers of physics and mathematics, primarily for secondary schools. Once more, she was the only female student in her class—to this day a difficult situation.

As it happened, among the handful of other students enrolling at the same time in the same section and for the same purpose, there was a bright young seventeen-year-old chap, about three and a half years her junior. He had also come to Switzerland from abroad, and also had already shown that he was devoted to the study of science. His name was, of course, Albert Einstein. Through their letters we can begin to glimpse the story of their love, their physics, and their other passions.

Young Einstein entered the Poly after a final year of secondary-school education at the Kantonsschule in Aarau, Switzerland. He had been there as a boarder in the home of the teacher of Greek and of history at the school, Jost Winteler and his wife, Pauline. They had seven children and kept a warm and comfortable house in which Einstein flourished. Einstein was soon as devoted to the Winteler parents as to his own, or perhaps even more. (Eventually, one of the Winteler sons married Einstein's sister, Maja, and one of the daughters married Einstein's closest friend, Michele Besso.) Moreover, while at Aarau, Albert and the youngest of the daughters, Marie, were greatly attracted to each other, and both sets of parents thought they would eventually make a good couple. In fact, if we look into the first volume of the published correspondence of Einstein (*The Collected Papers of Albert Einstein* [Princeton, N.J.: Princeton University Press, 1987]), one of the earliest letters (November 1896) is from Marie Winteler to him, starting with the appellation "Beloved Treasure" ("Geliebter Schatz") and ending with "In deep love, your Mariechen." It's a rather desperate love letter from one adolescent to another, in which she pleads with him not to break off the correspondence.

THE EINSTEIN ARCHIVE

Let me interpolate here to say something about the Einstein Archive that is the basis of much that follows. By the will of Albert Einstein, all correspondence, books, manuscripts, etc., in his possession at his death were eventually transferred to the Hebrew University Library in Jerusalem, and most of the correspondence is being published, several volumes out of an initially projected total of thirty having already appeared. The archive is a treasure trove, since Einstein corresponded not only with most of the major and many of the minor scientists of his time, but also with people in political life, from Gandhi and Roosevelt to Stalin; with writers and musicians, sages and cranks, and the wretched of this earth. In a way, the archive is a window on a good deal of the triumphs and terrors of the twentieth century, and it is now being used widely by scholars in the most diverse fields.

The fact that it exists in a form that can be readily used by scholars is in some part due to my stumbling on the documents almost by accident. After Einstein's death in 1955, I was asked to participate in a memorial symposium. To my surprise, I found that very little had been published by historians of physics on the impact of Einstein's work, and I thought that my contribution might be to begin such an assessment, based on whatever documents might be available. Philipp Frank, then a colleague in my department and the good friend and biographer of Einstein, recommended that I examine whatever manuscripts or letters I might find at the Princeton Institute for Advanced Study or at Einstein's old house on Mercer Street.

Arriving in Princeton, I was told that all available materials might be in some files in the big basement safe room at the institute's main building, Fuld Hall. I was not prepared for what I found there. In the near darkness, against the background of many tall metal file cabinets, sat Helen Dukas, Einstein's secretary since 1928. The gooseneck table lamp on her small desk provided the only illumination. She was busy, responding to letters relating to Einstein or requesting reprint rights. The whole scene reminded me of Juliet in the crypt. When I asked whether there were documents illustrating Einstein's way of thinking and working in physics, and his interaction with his colleagues, she opened some of the file drawers. Arranged in a way accessible to her but perhaps to no one else, here were the many thousands of letters, drafts of answers, published and unpublished manuscripts, etc., that had been carefully preserved by his devoted helpers over the years. After Einstein's immigration to the United States in 1933, most of the material then in existence had been spirited out of Nazi Germany, with the aid of the French Embassy in Berlin, and sent to the United States in a so-called diplomatic pouch arrangement.

Looking at this chaotic collection, and getting from Helen Dukas the distinct impression that more was stored at Einstein's house in Princeton in which she and his stepdaughter Margot were still living, I felt compelled to get this mountainous material organized into a workable archive. In those days, one of the large private foundations acted quickly and sensibly on requests for aid in cultural matters of this sort. So I was able to hire Helen and some physics graduate students from Princeton, and arranged for periodic visits, to help convert the files into an archive, to prepare a catalogue raisonné summarizing each of the documents, and to begin the search for copies of documents held by others. Eventually the archive turned out to contain some forty-five thousand documents, including those added since 1955 by gift or purchase. Among those are the copies of the fifty-four letters exchanged between Einstein and Marić between 1897 and 1903 (forty-three of his, eleven of hers), now published in the *Collected Papers*, which were made available in 1986 by the family of Einstein's first son, Hans Albert.

THE CORRESPONDENCE

Now back to the letters of the late 1890s. The first surviving one between Mileva and Albert is hers of 20 October 1897 (about a year after Marie's). At the Poly, she and Albert had been taking mostly the same required courses—although he also took classes in business, banking, and insurance statistics, to be on the safe side in case he should be needed to help run the family business (electrical engineering). But now, during her second year, Mileva had taken a leave for a few months to attend lectures in physics and mathematics at the University of Heidelberg. In that first mailing we have from her hand, she addresses the young man by the formal *Sie*, rather than the informal *Du*. Perhaps because of the ambiguity of just how to address him, her letter has neither initial nor final salutation. It is rather cool, apparently in reply to a four-page letter she had received from him awhile earlier, one of the many which have not been preserved. She writes that on a visit back home she had told her papa all about Albert: "You absolutely must

come [home] with me some day. You would enjoy yourself splendidly."
She adds, as if it were a non sequitur, that one of their mutual acquain-
tances seems to have become involved in a romance, and she concludes:
"What's the point of him falling in love nowadays, anyway? It's such an
old story. . . ." She finishes with a brief remark about a lecture she has
just heard on the kinetic theory of gases, based on the then-still controversial concept that gases consist of atoms.

Chronologically, the next document in the archives is an envelope
Albert had addressed to Mileva. But the envelope is empty—a reminder that as scholars or snoops we are restricted to whatever scraps
may be left over. Six weeks later, there is a letter from Albert to Mileva.
He also addresses her quite formally: "Geehrtes Fräulein." He looks
forward to her impending return to Zurich. Because she has fallen a
year behind her classmates there, he offers to make available his class
notes so that she will be able to catch up. (When she later used them,
she penned a correction of one of Einstein's small mistakes—in his
drawing of two vessels, she noted that his discussion should refer to the
vessel on the right, not the left—and this correction, properly credited
to her, has now appeared in Einstein's *Collected Papers*. The irony of it
will become clear later.)

Now the friendship is turning into a romance amid the experimental
equipment in the physics lab of the Poly. They spend much of their
time together studying for their courses, and doing the required experimental thesis project needed to obtain the diploma. During vacations,
Albert often goes back to visit his family in Milan, where his father has
tried to start a new business after an earlier failure of his company. On
those occasions we have letters between them. For example, Albert
writes in March 1899 that he realizes "how closely knit our psychological and physiological life is." He had shown a photograph of Mileva to
his mother, and writes: "Your photograph had a great effect on my old
lady"; she had studied it carefully, and "sends her greetings."

But Mother's greetings were not from the heart. Slowly Albert
realizes his parents' hostility to this friendship, and at the same time,
in reaction, he begins to see *them* in a new light. Early August 1899:
"My mother and sister seem a little petty and philistine to me, despite
the sympathy I feel for them. It is curious how gradually our lives

change us in the very subtle shadings of our soul, so that even the closest family ties dwindle into habitual friendship, and deep inside one becomes mutually so incomprehensible that one is in no way capable of empathizing with the emotions that move the other." And, again, in the second letter of that month: "I manage to escape their mindless prattle. . . . If only you were again with me for a while! We understand so well each other's dark souls, and also drinking coffee & eating sausages, etc. . . . . . ." (The six final periods are underlined.)

ELECTRODYNAMICS OF MOVING BODIES

Even as the romance is getting more stable and serious, another thing comes to the foreground. In virtually each of his longer letters, aside from personal matters often written in his witty and colloquial German to which translations can hardly do justice, Einstein goes into details of the physics that is preoccupying him—not merely rapturous accounts of what he is reading in his textbooks or in classics of science, but new ideas. For example, in that second letter of August 1899: "[I] am now rereading Hertz's "Propagation of Electric Force" with greatest care. The reason is that I did not understand Helmholtz's treatise. . . . I am more and more reaching the conviction that the electrodynamics of moving bodies, as it is presented today, doesn't correspond to reality, but can be presented more simply. The introduction of the term 'ether' into the theories of electricity has led to the conception of a medium of whose motion one can speak, but without, I believe, being able to connect a physical meaning with this way of speaking. I believe that electric forces are definable only for empty space, as Hertz also stressed." And so forth. This is six years before his major paper of 1905 on relativity theory, "On the Electrodynamics of Moving Bodies," containing in its title the very phrase he uses here, and which had also been used by Hertz. But it is perhaps revealing of their relative interests that while Einstein devotes much of his letters to his new ideas in physics, in Mileva's reply she is totally silent on that subject, and comments only on family matters and the like. On the whole, that asymmetry is true for most of their preserved exchanges. In the relatively few letters of Mileva's during the period under discussion, she never responds to Einstein's arguments about physics, nor does she otherwise write of scientific problems except to refer occasionally to a lecture she has heard or to a science book.

But Einstein continues to present his scientific ideas to her. Thus in his next letter (10 September 1899): "In Aarau [where he went to visit the Wintelers, probably with his sister Maja, who was entering the girls' school there] I had a good idea for investigating what effect a body's relative motion with respect to the luminiferous ether has on the velocity of propagation of light in transparent bodies." It is apparently a reference to some variant of the Fizeau experiment of 1851. As Einstein wrote much later, the clues provided by the Fizeau experiment and stellar aberration, together with the old experiment of Faraday on electromagnetic induction, "were enough" to put him on the trail of relativity theory.

Other tantalizing hints of his reading and thinking are sprinkled throughout his correspondence. For example, in a letter two weeks after the one just quoted, he writes to Mileva: "I also wrote to Professor Wien in Aachen about my work on the relative motion of the luminiferous ether against ponderable matter, the work which the 'boss' handled in such a negligent fashion." This presumably refers to his physics teacher, H. F. Weber. As one of Einstein's biographers, his son-in-law (Rudolf Kaiser, who published the book *Albert Einstein* under the pseudonym Anton Reiser in 1930) put it, during his student days, Einstein "wanted to construct an apparatus which would accurately measure the earth's movement against the ether," but was not able to do so because "the skepticism of his teachers was too great, the spirit of enterprise too small (p. 52)."

Einstein goes on to say: "I read a very interesting paper by Wien from 1898 on this subject." That paper, on theories of the motion of the presumed ether, described thirteen of the most important experiments trying to detect the earth's motion through that ether. The vast majority of them, ten, had negative results, and among those, indeed the last of them, was the Michelson-Morley experiment which later was elevated by some commentators as the only experiment of significance in the genesis of relativity theory—a position Einstein repeatedly disavowed.

In the same letter (28 September 1899): "I tend to brood too much—in short I see and feel the absence of your beneficent thumb under which the boundaries are kept." It is a puzzling and possibly revealing sentence, followed by other hints of his complex state of mind: "Don't worry about my now going so often to Aarau. The critical daughter with whom I was so madly in love four years ago [he is referring to Marie Winteler] is coming home. For the most part I feel quite secure in my high fortress of calm. But if I saw her again a few times, I would certainly go mad, that I know & fear like fire. . . . A thousand hearty greetings from your Albert."

As we turn to the letters of 1900, something new appears. They have agreed on an important step: to address each other finally by the familiar *Du* (thou), and to refer to each other by nicknames, part of the usual scenario of developing affections: "Mein liebstes Doxerl [doll]," but also "Meine süsse Kloane [little one]," "Meine liebe Miez [pussy-cat]," and variants on all of these. He in turn calls himself, and is often addressed by her as, Johannesl or Johanzel. In a letter to a friend, Mileva calls Einstein her "Hauptperson," her "significant other," as we now would put it.

By the end of July, a major turning point occurs for each of them, and therefore for their relationship. A letter from the university authorities of 27 July 1900 brings the news that Albert Einstein has passed his diploma examination. But Mileva, taking the examination at the same time, failed it. (Her grades were quite good in theoretical and practical physics, but not good enough in mathematics.) From other sources we know that, tired and depressed, Mileva decided to leave the Poly to visit her parents. A day or two later, Einstein wrote her from a summer vacation place in Switzerland where once again he had joined his parents and sister. It is a dramatic, even theatrical letter, which shows how his emotional life is rearranging itself. Let me quote it at length:

> I arrived in Sarnen the day before yesterday as planned. . . . We were met by Mama, Maja, and a carriage. . . . Maja took this opportunity to say [privately] that she had not dared to report anything about the "Dockerl" affair, and she asked me to "go easy" on Mama. . . . So we arrive home [at the hotel] and I go into Mama's room, only the two of us. First I must tell her about the exam [that he had passed, and Mileva failed], then she asks me quite innocently: "So, what will become of Dockerl now?" "My wife," I said just as innocently, but prepared for the expected "scene."

That followed immediately. Mama threw herself onto her bed, buried her head in the pillows, and wept like a child. After regaining her composure, she immediately shifted to a desperate attack: "You are ruining your future and barring your career. . . . No decent family will have her. . . . If she gets pregnant, you'll be in a mess." With this last outburst, I finally lost my patience. I vehemently denied the suspicion that we had been living in sin, and scolded her mightily. . . .

The next day things were better, largely because, as she said herself, "If they have not yet been intimate (which she had greatly feared) and are willing to wait longer, then ways and means can be found." The only thing that is out of the question for her is that we want to remain together always. The attempts at changing my mind came in expressions such as "Like you, she is a book—but you ought to have a wife." "By the time you are 30, she is an old witch," etc. . . . If only I could be with you again soon in Zurich, my little treasure! A thousand greetings and the biggest kisses from your Johannesl.

STUBBORN AS A MULE

He has now assumed the role as Mileva's defender and comforter (though one might doubt whether a letter like that would really soothe her). A few days later, he adds: "Mama and Papa are quite phlegmatic types and have less stubbornness in their entire bodies than I have in my little finger." On that he was probably quite right. When asked later why it was he who came to find the relativity theory, he gave two answers: that he had curiosity and the stubbornness of a mule, God's only gift to him; and that he continued to ask questions about the world that children eventually are taught not to ask.

The more Albert now experienced his family's rejection of Mileva, the more he was drawn toward her and away from them. He writes on 1 August 1900, "I long terribly for a letter from my beloved witch. I can hardly comprehend that we are for such a long time separated—only now do I see how terribly dear you are to me! Indulge yourself completely so you will become a radiant little darling and as mad as a street urchin [Gassenbub]." And two weeks later (14 August 1900), "How was I able to live alone before, my little everything? Without you I lack self-

*Einstein in about
1895-1896.*
(COURTESY OF
ELENA SANESI.)

*Mileva Marić in about
1896.*
(SCHWEIZERISCHE
LANDESBIBLIOTEK, BERN.)

Mileva's family in about 1883. Right to left: Mileva; her father, Miloš; his adopted daughter Nana and wife, Marija, with baby daughter Zorka. (FROM DESANKA TRBUHOVIĆ-GJURIĆ, IM SCHATTEN ALBERT EINSTEINS, PAUL HAUPT VERLAG, BERN, 1988; REPRODUCED WITH PERMISSION OF THE PUBLISHER.)

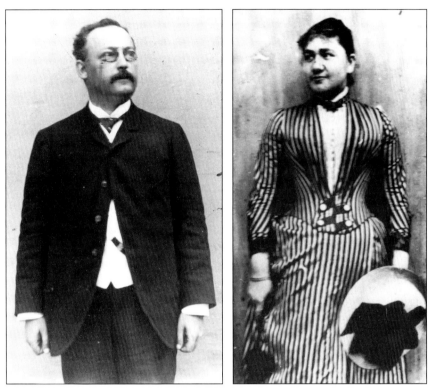

Einstein's parents, Hermann and Pauline (née Koch). (EINSTEIN ARCHIVE, HEBREW UNIVERSITY, JERUSALEM.)

Wedding portrait of Mileva and Albert in Bern, January 1903.

A Love Letter
The letter from Mileva to Albert whose first page is shown at left was written in 1901, probably soon after the Como expedition. The translation cannot do justice to the endearing diminutives (Schatzerl, Brieferl, Weiberl, Manderl, *even* Gotterl), *which are reminiscent of the lovers Papageno and Papagena in Mozart's* Magic Flute. *The following is a partial translation:*

Dearest Little Treasure!
Now I've also received your second little letter and I'm so happy, beyond all measure. How dear you are, oh how I'll kiss [busseln] you, I can hardly wait for the end of the week, when you come. . . . If you come Saturday you might be able to sleep at our [rooming house], because someone is leaving Friday. I'll ask (the landlady), she'll do it for me if it's possible. Till then I'll work very hard so I'll be quite free to enjoy your company—God [Gotterl], how beautiful the world will look when I'm your little wife, you'll see there'll be none happier in the whole world, and then it must be the same for the man [Manderl].

Stay well my sweet little treasure and come in good cheer at week's end to
Your woman [Weiberl].

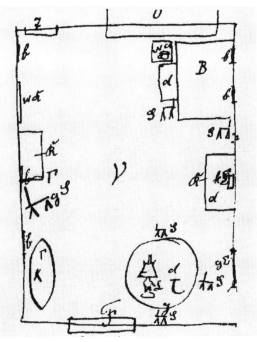

A sketch. In February 1902 Albert send Mileva this sketch he had made of his furnished room in Bern. His explanation (below) of the symbols is full of diminutives and dialect expressions, as is so much of their correspondence. The last item (Γ) is perhaps a picture of Mileva. (COURTESY OF ALBERT EINSTEIN ESTATE.)

B = Betterl *[little bed]*
b = Bilderl *[little picture]*
d = Deckerl *[little mat]*
gS = grossartiger Sessel *[terrific easychair]*
$g\Sigma$ = grossartiger Spiegel *[terrific mirror]*
J = Johonzel *[Little Johnny (himself)]*
K = Kasten *[chest]*
k = Kanapyen *[sofa]*
$k\Sigma$ = kleiner Spiegel *[small mirror]*
N = Nacht-Topf & Tisch *[chamberpot & table]*
F = Fensterl *[little window]*
O = Ofen *[stove]*
S = Stühlerl *[little stool]*
T = Thüre *[door]*
τ = Tisch *[table]*
ν = nichts *[nothing]*
U = Uhrerl *[little clock]*
Γ = Gell da schaugst! *[I'll bet you're surprised!]*

Mileva and Albert Einstein with their son Hans Albert in 1904. (EINSTEIN ARCHIVE, HEBREW UNIVERSITY, JERUSALEM.)

Einstein and his son Hans Albert in 1904 or early 1905.
(© EVELYN EINSTEIN.)

Mileva Einstein and her sons, Eduard (left) and Hans Albert, around 1914.
(EINSTEIN ARCHIVE, HEBREW UNIVERSITY, JERUSALEM.)

Hans Albert and Eduard (right) with Einstein in the mid-1920s.
(© EVELYN EINSTEIN.)

confidence, passion for work, and enjoyment of life—in short, without you my life is no life."

But as one would expect, the escape from the merely personal is always at hand for him. On 30 August 1900: "My only diversion is studying, which I am pursuing with redoubled effort, and my only hope is you, my dear, faithful soul." And 13 September 1900: "Boltzmann is quite magnificent. . . . He is a masterful writer. I am firmly convinced of the correctness of the principle of the theory, i.e., I am convinced that in the case of gases we are really dealing with discrete mass points of definite, finite magnitude that move according to certain conditions." This is of course a basis for his paper, published five years later, with the proof of the molecular hypothesis through Brownian motion.

On 19 September 1900, Einstein shows once more he is anxious to escape a "philistine" existence—in his wide reading he was drawn to Schopenhauer, who for young people at the time was a strong voice against following the crowd, and in later letters Einstein often referred to himself as a gypsy at heart. Now he paints an optimistic picture for the two young people in love (19 September 1900): "No matter what may happen we'll have the most delightful life in the world. Pleasant work and being together—and what's more, we both are now our own masters and stand on our own two feet and can enjoy our youth to the utmost. Who could have it any better? When we have saved up enough money, we shall buy ourselves bicycles and take a bike tour every couple of weeks."

"OUR WORK ON RELATIVE MOTION"

By late October 1900, Einstein must have rejoined Mileva in Zurich, and there are no more letters between them until spring 1901, when he visits his parents again. She has been left alone, studying to take that examination for the second time. It is a low point for her; perhaps it is dawning on her that she will fail again and will never get her degree. And at that moment he writes a letter (27 March 1901) from that hostile camp which contains a phrase on which some commentators fastened for a time. "My dear kitten. Many thanks for your letters and for all true love contained therein. I kiss you and hug you with all my heart for it. . . . I'll try to get an assistantship in Italy. First of all that removes a major problem, namely the anti-Semitism that, in the German lands, would be as distasteful to me as it is obstructive. . . . You are and will remain a shrine for me to which no one has access; I also know that of all people you love me the most and understand me the best. I assure you that no one here would dare or even want to say anything bad about you. I'll be so happy and proud when we together shall have brought our work on relative motion victoriously to a conclusion!"

"*Our* work on relative motion." In their surviving correspondence, it is his only use of the word *our* in the context of the early labors that led to the relativity theory four years later; but it will occur again in other contexts, and requires some comments later to understand what may have been meant.

We come now to another key event in the life of these struggling young people. Now that Einstein had his diploma, he was hoping to find a job. His applications for assistantships to various physics professors in Switzerland, Germany, and Italy yielded no encouraging replies, but in mid-April 1901 he had good news. He was invited to be a substitute teacher for two months at the Technical High School in Winterthur, and also there was a letter from his friend Marcel Grossmann, saying that (thanks to the intervention of Marcel's father) Einstein might perhaps have a chance to get a position at the Swiss patent office in Bern. Strangely, his report to Mileva on all this is given only a few lines at the beginning of his letter to her. The main topic is a lengthy discussion of what he calls his new "wonderful idea" to apply a theory of molecular forces to gases, followed by a long exposition of the physics involved.

A few days later, still in high excitement, he again tries to lift Mileva's darkening mood—she had even begun to look for a secondary school position in Zagreb, and incidentally, she was also jealous of the attention Einstein seems to have paid to some of her women friends. So on 30 April 1901, Einstein writes her: "My dearest little child [Kinderl]! I just don't let up. You absolutely must come to me in Como [in Northern Italy], you sweet little witch. . . . And I love you so much again! It was only out of nervousness that I was so mean to you. You will hardly recognize me now that I have become so bright and cheerful, and I'm longing so much to see again my dearest Doxerl. . . . Come to me in Como and bring my blue morningrobe in which we can wrap ourselves up, and don't forget your binoculars." To be sure, this suggestion is immediately followed by his opinions on Boltzmann's gas theory. He believes that there is now on this subject "enough empirical material for our investigation," to be found in a book by O. E. Meyer, and asks her to look for it the next time she is in the library.

There is something a little frantic in this letter. Obviously, life is now looking quite different from their respective positions, and he is trying to cheer her up and to bind her to him. After some hesitation, Mileva accepts his invitation. She says she has just "received a letter from home today that has made me lose all desire, not only for having fun, but for life itself."

That excursion, later described in one of her letters to her friend Helene Savić, turned out to be perhaps the most important threshold point for the couple. They went north from spring-bound Como to the Splügen Pass, some sixty-five hundred feet high, where the snow was in places still nearly twenty feet deep. They rented a sled and then went further on foot while more snow was falling. Mileva wrote later to a friend that she held on to him firmly throughout. It was a great success. "How happy I was again to have my darling for myself a little, especially because I saw that he was equally happy." It was all the more glorious for them to be together because for the next two months Einstein would be in Winterthur on his temporary teaching job.

"How Is the Boy?"

But perhaps the trip had gone all too well. A few weeks later (end of May 1901), one of Einstein's surviving letters to Mileva begins with his typical account of his almost physical euphoria on encountering in his reading a new physics problem—in this case the photoelectric effect—but then he almost seamlessly continues to write about the main new fact that has entered their lives: "I just read a wonderful paper by Lenard on the generation of cathode rays by ultraviolet light. [We know that Einstein's following up on that experiment was responsible for the earliest of his major 1905 papers, the one usually referred to as his photon theory of light, for which he was awarded the Nobel Prize in 1922.] Under the influence of this beautiful piece, I am filled with such happiness and joy that I absolutely must share some of it with you. Be of good cheer, dear, and don't fret. After all, I am not leaving you, and will bring everything to a happy conclusion. One just has to be patient! You'll see that one does not rest badly in my arms, even if things are beginning a little stupidly. How are you, darling? *How is the boy?*"

Mileva had discovered she was pregnant. This was surely not the first time that they had talked about it—he had visited her in Zurich sometimes during his work in Winterthur. But little did he seem to realize how this pregnancy would change their relationship. His first scientific paper, on capillarity, had been sent off the previous December. He was now launched as a publishing scientist (even though he thought later that his first published papers were quite insignificant). But for Mileva, becoming a science school teacher (not to speak of obtaining a university position of which she may only have dreamed) had already become very questionable. She was still studying for her second try at the examination in the teacher-track section of the Poly, which she would bravely take at the end of July 1901, and, three months pregnant, fail with the same low total mark as before. And bearing an illegitimate child was of course in those days considered to be deeply shameful.

But at that point, in late spring 1901, Albert, despite his un-quenchable optimism, was not showing much more promise—a twenty-two-year-old still without firm employment. His eventual job, a provisional appointment as Technical Expert Third Class at the Patent Office, actually did not materialize until about a year later. With little likelihood for even a menial permanent post, and with no moral sup-port for his liaison from either his family or hers (there are even references to angry letters from Einstein's parents to Mileva's, which caused predictable sorts of rows), he now faced, in straitlaced Switzer-land, the prospect of becoming the impoverished father of an illegiti-mate child.

Here was a test of his fundamental morality. What would many a normal young man have done at that time under these conditions? For Einstein the answer was clear. A letter of 7 July 1901 to Mileva goes like this: "Rejoice now in the irrevocable decision I have made! About our future I have decided the following: I'll look for a position *immediately*, no matter how modest. [In fact, he seems to be thinking about going into an insurance firm as a clerk, or making do by giving hourly lessons.] My scientific goals and personal vanity will not prevent me from accepting the most subordinate role. As soon as I have such a position I will marry you and take you to live with me, without writing a word of it to anyone until everything has been settled. Then no one can cast a stone upon your dear head, and woe unto him who dares to set himself against you. When your parents and mine are presented with a *fait accompli*, they'll just have to reconcile themselves to it as best they can."

For whatever reason, this is not how it turned out. By November 1901, they write each other again, because Mileva has gone home to her parents where she plans to have their baby. Albert is in yet another lowly temporary teaching position, in Schaffhausen, from which he soon manages to get himself dismissed for impudent behavior (such as asking for a better salary). Mileva writes on 13 November 1901 to suggest that they keep the news of the pregnancy quiet: "I believe we should say nothing yet about Lieserl [to other friends]." The unborn child has at least been given a name; and it turned out later that the mother had guessed the gender correctly, although Albert replies, on 12 December 1901, that: "Secretly [I] prefer to imagine a Hanserl." He hopes the Bern job will come through, but even if it does not, then "together, we'd surely be the happiest people on earth. We'll be students as long as we live and won't give a damn about the world. . . . The only problem that still needs to be solved is the question of how we can take our Lieserl to us; I do not want us to have to give her up. Ask your Papa, he's an experienced man. . . ." And he adds, "She shouldn't be stuffed with cow's milk because it might make her stupid. Yours would be more nourishing, right?" And next, almost predictably: "I have had another rather obvious but important scientific idea about molecular forces. . . ."

"So We Don't Become Old Philistines"

Five days later, on 17 December 1901: "I am busily at work on an electrodynamics of moving bodies, which promises to be a capital piece of work. I wrote to you that I doubted the correctness of the idea about relative motion. But my reservations were based on a simple calculation error. Now I believe in it more than ever." And further on, "Give your mother my best greetings and tell her I am looking forward to the spanking with which she will do me honor some day." Two days later, 19 December, "Sorry, I forgot your birthday once again," and more thoughts on his electrodynamics. Nine days later, December 28, "When you become my dear little wife, we'll diligently work on science together so we don't become old philistines, right?"

Lieserl was born in January 1902, perhaps at Mileva's home in Novi Sad. Soon after he heard the news, Albert writes her (4 February 1902, Bern): "My beloved treasure! Poor dear sweetheart, what you have to suffer that you now can't even write to me yourself. It is such a shame that our dear Lieserl must be introduced to the world this way! I hope you are bright and cheerful by the time my letter arrives. . . . Now you see that it really turned out to be a Lieserl, just as you had wished. Is she healthy? Does she cry properly? What are her eyes like? Which one of us does she resemble more? Who is giving her milk? Is she hungry? . . . Couldn't you have a photograph made of her when you are again in good health? Can she turn her eyes in different directions to see things? You can now make observations. I'd like to have a Lieserl myself some day—it must be fascinating. She is certainly able to cry already, but won't learn how to laugh until much later. Therein lies a profound truth. When you feel a little better you will have to draw a picture of her." He has placed an advertisement in the local newspaper, offering to give private lessons.

"What Has Befallen Lieserl"

Now briefly the rest of the story of this star-crossed couple's early years. The marriage took place on 6 January 1903. Lieserl, whose existence was unknown to all Einstein scholars until these letters were found and published in 1987, is mentioned in only one other letter of Albert, on 19 September 1903, when Mileva is back at her family's home, as he puts it, "to hatch a new chick"—that is, to give birth to the first of their two sons, Hans Albert. Einstein writes, "I am very sorry about what has befallen Lieserl. It is so easy to suffer lasting effects from scarlet fever. If only this will pass safely. As what is the child registered? We must be very careful that problems don't arise for her later." The Einstein Collected Papers Project mounted an extensive search to trace her through birth registers and even gravestones; but nothing has been found. If she survived, very likely she was adopted by one of Mileva's relatives, in accord with custom in those days for such cases. Although now quite improbable, it is of course technically still possible that she might surface, saying, "My name is Anastasia, and I want the family papers."

This is not just idle speculation. Something like this nearly happened. In the 1930s, after Einstein had moved to America, a woman turned up in Germany who claimed to be a long-lost daughter of Einstein's. Her story persuaded a number of Einstein's friends in Europe, who conveyed their surprise and dismay to him. Eventually it turned out that she was a fraud, an actress born in Vienna in 1894, when Einstein was fifteen. That episode illustrates one of the many cases in which the mature Einstein, like other highly visible persons, became a target of con artists and sensationalizers, and, in part, explains why in Einstein's later years those near him tended to seem overprotective.

But there is no doubt that the birth of Lieserl took a heavy toll on Mileva herself, whose attitudes and behavior tended to become more and more brooding after those secret events. Perhaps she was blaming herself for the decision to give up her daughter, or blamed Albert for acquiescing in it, or possibly even for urging her to acquiesce. We simply do not know. At any rate, the periods of melancholy that increasingly enveloped Mileva became in time evident to everyone. Einstein later described her as taciturn, suspicious, and depressed, and attributed it to a background of schizophrenia on her mother's side. Albert's and Mileva's younger son, Eduard, also showed signs of the illness. More and more, the dark side of life began to show itself to the family.

But at least the first few years of marriage seem to have been pleasant enough on the whole. Shortly after the wedding, Albert wrote to his good friend Michele Besso (January 1903): "Well, now I am a thoroughly married husband and lead a nice comfortable life with my wife. She takes excellent care of everything, cooks well, and is always in a good mood." Similarly, Mileva wrote on 20 March 1903 to her confidante, Helene Savić: "I am, if possible, even more attached to my dear treasure than I was in the Zurich days. He is my only companion in society, and I am happiest when he is beside me."

BABY CARRIAGE AND CHEAP CIGARS

A family friend has described the couple's life after the birth of Hans Albert in May 1904, during Einstein's most productive phase in his Swiss years. Mileva is out on some errand, and Albert is sitting at the kitchen table, one hand pushing the baby carriage back and forth, while the other is writing one of his world-shaking papers. The room is enveloped in the acrid smoke from one of the cheap cigars which were then his luxury. But what is ominously missing from all these comfortable images of the early years of marriage is any evidence of the kind of life they planned when first drawn to each other—two schoolteachers coming home after work for an evening of studying together happily, and perhaps writing something new for the major physics journals, as a few such teachers then still could hope to do.

The rest was also not what they had dreamed of in their early letters. By 1909, Einstein's reputation was rising fast in science, despite the fact that in his style of thinking and publications, as well as in his independent personal demeanor, he seemed to thumb his nose at authority of every kind. But the marriage was clearly in trouble, as seen by letters to friends. She wrote (again to Savić): "You see, with such fame [for Einstein] not much time remains for his wife. . . . The pearls are given to the one, the other gets just the shell. . . . You see, I am very starved for love." All traces of her interest in a career in science or in teaching were gone. Her friends during those years describe Mileva's brooding, dark moods, introversion, long periods of silence even at home, extreme taciturnity, unresponsiveness, melancholia. The happy years of love between Albert and Mileva had ended. In 1911, the family moved to Prague where Einstein served for about a year as a university professor. But there the family felt uncomfortable. In 1912, they returned to Zurich, where he had been offered a professorship at their old school, the Poly.

Shortly before World War I, Albert accepted the call to the distinguished position as member of the Prussian Academy of Sciences and research professor at the Kaiser Wilhelm Institute in Berlin. For him, living in the country he had chosen to leave as a youngster was hard enough, but Mileva found it unbearable. She packed up the two boys and took them back to Switzerland in 1914. During the following few years, while Einstein persevered through bouts of illness in almost superhuman effort to fashion the general relativity theory, he was also courted by a cousin in Berlin, Elsa Löwenthal, with whom he had an occasional flirtation starting in 1912. In 1919, Mileva divorced Albert and continued to care for her children as best she could. Hans Albert went on to become a professor of engineering at the University of California, Berkeley. But the younger son, Eduard, was in and out of psychiatric clinics, treated for schizophrenia, and died in 1965 in the same Swiss clinic in which his mother's sister Zorka had also been treated for mental instability.[4] Soon after the divorce, Elsa Löwenthal and Albert were married. When Einstein received the Nobel Prize in 1922—which knowledgeable physicists had predicted would come sooner or later—he transferred the prize money to Mileva, as he had promised in the divorce proceedings. In 1948, in poor physical and mental condition, Mileva died in Switzerland.

WHAT HAVE WE LEARNED?

So, what have we learned? Some main points would include these:

1. Thanks to these letters, we know of course much more about Mileva Marić, her courage, hopes, early ambition, and disappointments. She left no evidence of originality as a future major scientist— but that is of course true for most who aspire to such a career, whether male or female. During the early, good years she and Albert longed palpably for each other's companionship, and each gained emotionally and psychologically from it. Einstein also valued her intellectually, and not only because he, an autodidact throughout his life, always needed someone who understood him, to talk with about new ideas (as did Niels Bohr). One remembers the warning the physicist Max von Laue gave a friend in 1912 in preparing him for his first meeting with Einstein: "You should be careful that Einstein doesn't talk you to death. He loves to do that, you know."

We also know more about Mileva's increasing demoralization or loss of self-esteem—caused perhaps by the loss of Lieserl, or by the increasing fame of her husband, with all the demands this entailed, or by their disappointment with each other, when the original project of a cozy couple studying science together came to naught.

2. And we know more about the young Albert Einstein. In the years I have covered, he was of course not the sad-eyed icon and patriarch of late years, but a lusty and irrepressible young man who attempted to combine his emotional and intellectual passions. This is symbolized for me by the fact that some of his early letters to Mileva were written on pages torn from his notebooks, the back of the sheet containing calculations from the theory of electromagnetism. I recall here the answer Erik Erikson said he received from Freud when Erikson asked what might be the primary components of a good life. Freud had put it memorably in three words: *Arbeit und Liebe* (work and love).

We already knew that in his later years Einstein was never discouraged by failures in his doomed, late project to find a unified field theory; but here we also saw that during the decade we have covered practically nothing could squelch Einstein's fundamental optimism, either about his work in science or about his faith in Mileva.

3. As I noted, during the years when Albert and Mileva were passionately attached to each other, and chiefly when she seemed in need of psychological support, he occasionally used in his letters to her such phrases as *our work* or *our investigation*. Starting in 1990, a small number of writers have attempted to inflate the potential meaning of these phrases to include the possibility that Mileva was actually responsible for either the physics or the mathematics in Einstein's published relativity paper of 1905, but was deprived of any credit for it. For a time this allegation caused extensive and sensationalistic press coverage around the world, even though one of its two main promoters revealed his purpose in an interview (*New York Times*, 27 March 1990): "My point is to say that the king had no clothes." From time to time the story rears its head again. But careful analysis of the matter by established scholars in the history of physics, including John Stachel, Jürgen Renn, Robert Schulmann, and Abraham Pais, has shown that scientific

collaboration between the couple was minimal and one-sided. Einstein's occasional use of the word *our* was chiefly meant to serve the emotional needs of the moment. To quote Stachel's essay: "To sum up, Marić seems to have encouraged and helped Einstein in a number of ways during their years together, notably as the alter ego, to whom he could express his ideas freely while developing them in isolation from the physics community. She also appears to have helped him by looking up data, suggesting proofs, checking calculations, and copying some of his notes and manuscripts. He never publicly acknowledged this help [nor did she claim it in her letters to him or to anyone else], nor did a truly creative collaboration ever develop."[5]

Ironically, the exaggeration of Mileva's scientific role, far beyond what she herself ever claimed or could be proved, only detracts both from her real and significant place in history and from the tragic unfulfillment of her early hopes and promise. For she was one of the pioneers in the movement to bring women into science, even if she did not reap its benefits. At great personal sacrifice, as it later turned out, she seems to have been essential to Albert during the onerous years of his most creative early period, not only as an anchor of his emotional life, but also as a sympathetic companion with whom he could sound out his highly unconventional ideas during the years when he was undergoing the quite unexpected, rapid metamorphosis from eager student to first-rank scientist.

4. From these letters we now know more of Einstein's steps toward his major works of 1905 and beyond: what he read and when (e.g., Wien), and how long it took him to go through the early, formative stages of what were to become the papers on relativity, Brownian movement, and the quantum theory of light.

5. We have also relearned the old lesson that establishing and persisting in a long-term loving relationship was for Albert and Mileva, as it is for so many, even more difficult than working on the awesome scientific problems which Einstein knew how to solve. We can never claim to know what it felt like to be them. But their attempts during the first few years to include each other in their life and their work—to make it one life—might be seen in terms of a process of the loosening and expanding of the boundary of one's ego so as to merge with each

other—not far from the Platonic theory of love as the enhancement or expansion of the self when it is completed by merging with its cognate partner. Or, to put it better, in the words of Shakespeare's Portia: "One half of me is yours, the other half yours, Mine own I would say; but if mine, then yours, And so all yours!" In this particular case, the initial expectations of making it *one* life were evidently quite unrealistic, and each side eventually perceived good reasons for feeling deeply disappointed. At any rate, when Einstein's second marriage (to Elsa) also proved unsatisfactory, Einstein accused himself wistfully of failing rather ignominiously in two attempts.

6. And finally, we have learned that some mysteries remain more or less as they were before. The surviving letters are like fragments of pottery from which one might guess at an ancient civilization. On the personal side, neither of the two people can be fully comprehended at this distance. On the scientific side, these letters do not help explain how a barely employable young man, with only the formal training of a schoolteacher, could rearrange the structure of physics set up by such giants as Maxwell, Boltzmann, Lorentz, and Planck. Perhaps (as proposed by Renn and Schulmann) Einstein was helped greatly by approaching physics first through reading, as an adolescent, a popular work on natural science, written in the middle of the nineteenth century by Aaron Bernstein. In his "Autobiographical Notes," Einstein refers to it as one of the books he had the good fortune of hitting upon, which were "not too particular in their logical rigor, but which made up for this by permitting the main thoughts to stand out clearly and synoptically. . . . [Particularly Bernstein's set of volumes] limited itself almost throughout to qualitative aspects . . . a work which I read with breathless attention." In this way he gained early a broad conceptual overview of physics on which to build and add technical details. That overview allowed him to draw connections between pedagogically separate subjects when he encountered them in university courses. In short, Einstein started his formal training with a preference for seeing phenomena in the context of the unity of nature, as drawn from Bernstein's writings and the widely-read *Kosmos* of Alexander von Humboldt.

One might conclude that Einstein began with the outlines of a valid scientific worldview rather than with the usual student's concern to master an encyclopedia of individual subjects. This is one more example of Einstein's position as an outsider, a "marginal" person in science, not beholden to any of the existing theories, and hence free to let his driving curiosity and overflowing originality create new ones—even while he and Mileva Marić, against all the obstacles and prejudices, were trying to put their passions for science and for each other into the service of forging a life together.[6]

Chapter 9

"WHAT, PRECISELY, IS THINKING?"... EINSTEIN'S ANSWER

How DID Albert Einstein do his thinking? At first glance an answer seems impossible. His work was carried out at the very frontiers of physics and of human ability. And his mind was not open to easy study from the outside, even by those who worked with him—as was discovered by the physicist Banesh Hoffmann who, with Leopold Infeld, was Einstein's assistant in 1937. Hoffmann has given an account of what it was like when he and Infeld, having come to an impassable obstacle in their work, would seek out Einstein's help. At such a point, Hoffmann related,

> We would all pause and Einstein would stand up quietly and say, in his quaint English, "I will a little think." So saying, he would pace up and down and walk around in circles, all the time twirling a lock of his long grey hair around his forefinger. At these moments of high drama, Infeld and I would remain completely still, not daring to move or make a sound, lest we interrupt his train of thought.

Many minutes would pass this way, and then, all of a sudden,

> Einstein would visibly relax and a smile would light up his face ... then he would tell us the solution to the problem, and almost

always the solution worked.... The solution sometimes was so simple we could have kicked ourselves for not having been able to think of it by ourselves. But that magic was performed invisibly in the recesses of Einstein's mind, by a process that we could not fathom. From this point of view the whole thing was completely frustrating.[1]

But if not accessible from the outside, Einstein's mind was accessible from the inside, because like many of the best scientists, he was interested in the way the scientific imagination works, and wrote about it frankly. As far as possible, we shall follow the description, quite accessible and in his own words, of how he wrestled with theories of fundamental importance. Needless to say, we shall not be under any illusion that by doing so we can imitate or even fully "explain" his detailed thought processes, nor will we forget that other scientists have other styles. But Einstein's humane and thoughtful description of scientific reasoning will serve as a reminder of how false the popular, hostile caricatures are that depict contemporary scientific thought, as analyzed in Chapters 1 and 2.

There are numerous sources to draw on, for Einstein wrote about his view of the nature of scientific discovery, in a generally consistent way, on many occasions, notably in the essays collected in the book *Ideas and Opinions* and in his letters. He was also intrigued enough by this problem to discuss it with researchers into the psychology of scientific ideas and with philosophers of science. Indeed, from his earliest student days, Einstein was deeply interested in the theory of knowledge (epistemology). He wrote, "The reciprocal relationship of epistemology and science is of noteworthy kind. They are dependent upon each other. Epistemology without contact with science becomes an empty scheme. Science without epistemology is—insofar as it is thinkable at all—primitive and muddled."[2]

There are two especially suitable routes to Einstein's thoughts. One is a set of pages near the beginning of the "Autobiographical Notes," which he wrote in 1946.[3] It is the only serious autobiographical essay he ever wrote, and he called it jokingly his own "obituary." It gives a fascinating picture of Einstein's contributions as he viewed them, looking back at the age of sixty-seven. The essay is chiefly an account of his intellectual development rather than an autobiography in the usual sense. We shall now use this remarkable document to learn from his own words, while avoiding the use of technical, philosophical terminology, as he himself avoided it. All quotations not otherwise identified are from the pages of this text.

The other path to an understanding of Einstein's way of thinking is found in some letters he wrote to an old friend after publication of the "Autobiographical Notes." These allow Einstein to rebut, so to speak, a few of the objections a reader of the autobiographical essay might have, and I discuss them therefore at the end of this chapter, where Einstein should have the final word.

THE COURAGE TO THINK

It certainly is curious to start one's autobiography, not with where and when one was born, the names of one's parents, and similar personal details, but to focus instead on a question which Einstein phrases simply: "What, precisely, is thinking?" Einstein explains why he has to start his "obituary" in this way: "For the essential in the being of a man of my type lies precisely in *what* he thinks and *how* he thinks, not in what he does or suffers."[4]

From this viewpoint, thinking is not a joy or a chore added to the daily existence. It is the essence of a person's very being, and the tool by which the transient sorrows, the primitive forms of feeling, and what he calls the other "merely personal" parts of existence can be mastered. For it is through such thought that one can lift oneself up to a level where one can think about "great, eternal riddles." It is a "liberation" which can yield inner freedom and security. When the mind grasps the "extra-personal" part of the world—that part which is not tied to shifting desires and moods—it gains knowledge which all men and women can share regardless of individual conditions, customs, and other differences.

This, of course, is precisely why the laws of nature, toward which these thoughts can be directed, are so powerful: their applicability in principle can be demonstrated by anyone, anywhere, at any time. The laws of nature are utterly shareable. Insofar as the conclusions are right, the laws discovered by a scientist are equally valid for different thinkers, or *invariant* with respect to the individual personal situations. Einstein's interest in this matter seems to be not unrelated to his work in the physics of relativity: The essence of relativity theory is precisely that it provides a tool for expressing the laws of nature in such a manner that they are invariant with respect to differently moving observers.

As his "Autobiographical Notes" show, Einstein was also aware that life cannot be all thought, that even the enjoyment of thought can be carried to a point where it may be "at the cost of other sides" of one's personality. But the danger which more ordinary persons face is not that they will abandon their very necessary personal ties, but that the society surrounding them will not say often enough what Einstein here suggests to his wide audience: that the purpose of thinking is more than merely solving problems and puzzles. It is instead, and most importantly, the necessary tool for permitting one's intellectual talent to come through, so that "gradually the major interest disengages itself . . . from the momentary and merely personal." Here Einstein is saying: Have the courage to take your own thoughts seriously, for they will shape you. And significantly, Einstein meant his whole analysis to apply to thinking on any topic, not only on scientific matters.

THINKING WITH IMAGES

Having touched on the *why* of thinking, the autobiography takes up the *how* of thinking and strangely seems to be concerned with "pictures" (*Bilder*):

> What, precisely, is "thinking"? When, at the reception of sense-impressions, memory-pictures emerge, this is not yet "thinking." And when such pictures form series, each member of which calls forth another, this too is not yet "thinking." When, however, a certain picture turns up in many such series, then—precisely through such return—it becomes an ordering element for such series, in that it connects series which in themselves are unconnected. Such an element becomes an instrument, a concept.

Adhering to one of several contesting traditions in psychology and philosophy and perhaps particularly influenced by Helmholtz and

Boltzmann, Einstein holds that the repeated encounter with images (such as "memory pictures") in a different context leads to the formation of "concepts." Thus, a small child might form the concept "glass" when he or she experiences that a variety of differently shaped solids are hard, transparent, and break on being dropped.

A concept must of course eventually be put into a form where it can be communicated to others; but for private thought it is not necessary to wait for this stage. For some people, including such physicists as Faraday and Rutherford, the most important part of thinking may occur without the use of words. Einstein writes: "I have no doubts but that our thinking goes on for the most part without use of signs (words) and beyond that to a considerable degree unconsciously." Such persons tend to think in terms of images to which words may or may not be assignable; Einstein tells of his pleasure in discovering, as a boy, his skill in contemplating relationships among geometrical "objects"—triangles and other nonverbal elements of the imagination. As we saw in Chapter 4, Einstein explained in a letter to the mathematician Jacques Hadamar that in his thinking he used not words but "certain signs and more or less clear images which can be voluntarily produced and combined." Einstein's letter continued as follows:

> The psychical entities which seem to serve as elements in thought are certain signs and more or less clear images which can be "voluntarily" reproduced and combined. . . . But taken from a psychological viewpoint, this combinatory play seems to be the essential feature in productive thought—before there is any connection in words or other kinds of signs which can be communicated to others. The above-mentioned elements are, in my case, of visual and some muscular type. Conventional words or other signs have to be sought for laboriously only in a secondary stage, when the mentioned associative play is sufficiently established and can be reproduced at will.[5]

Einstein's ability to visualize is evident in the brilliant use he made of "thought experiments" (*Gedankenexperimente*). His first came to him at the age of about sixteen, when he tried to imagine that he was pursuing a beam of light and wondered what the observable values of the electric and magnetic field vectors would be in the electro-magnetic wave making up the light beam. For example, looking back along the beam over the space of one whole wavelength, one should see that the local magnitudes of the electric and magnetic field vectors increase point by point from, say, zero to full strength, and then decrease again to zero, one wavelength away. This seemed to him a paradoxical conclusion. Already at that age, he seems to have assumed that Maxwell's equations must remain unchanged in form for the ob-server moving along the beam; but from those equations one did not expect to find such a stationary oscillatory pattern of electric and magnetic field vectors in free space. He realized later that in this problem "the germ of the special relativity theory was contained." (Among other examples of visualized *Gedankenexperimente*, Einstein related one which he said had led him to the general theory of relativity—as noted in Chapter 4.)

THE FREE PLAY WITH CONCEPTS

Having stressed the role of images and memory pictures, including *Gedankenexperimente*, in thinking, and having defined "concepts" as the crystallized products, the unvarying elements found to be common to many series of such memory pictures, Einstein now makes a startling assertion: "All our thinking is of the nature of a free play with con-cepts." This sentence has to be unraveled for it deals with two opposite but equally indispensable elements in all human thought, the empirical and the rational.

Even if one grants that "free play" is still play within some set of rules—similar to tentatively trying out a word to see if it fits into a crossword puzzle—by no means all philosophers would agree with Einstein's position. Some would argue that the external world imposes itself strongly on us and gives us little leeway for play, let alone for choosing the rules of the game. In Einstein's youth, most of his contemporaries believed in Immanuel Kant's description of the boundaries of such "play," namely, that they were to be fixed by two intuitions which are present in one's mind already at birth (i.e., *a priori*): Newtonian absolute space and absolute time. Only a few disagreed, including Ernst Mach, who called absolute space "a conceptual monstrosity, purely a thought-thing which cannot be pointed to in experience."

Thus Einstein was struggling anew with the old question: What precisely is the relation between our knowledge and the sensory raw material, "the only source of our knowledge"?[6] If we could be sure that there is one unchanging, external, "objective" world that is connected to our brains and our sensations in a reliable, causal way, then pure thought can lead to truths about physical science. But since we cannot be certain of this, how can we avoid falling constantly into error or fantasy? David Hume had shown that "habit may lead us to belief and expectation but not to the knowledge, and still less to the understanding, of lawful relations."[7] Einstein concluded that, "In error are those theorists who believe that theory comes inductively from experience."[8]

In fact, he was skeptical about both of the major opposing philosophies. He wrote that there is an "aristocratic illusion [of subjectivism or idealism] concerning the unlimited penetrating power of thought," just as there is a "plebian illusion of naive realism, according to which things are as they are perceived by us through our senses."[9] Einstein held that there is no "real world" which one can access directly—the whole concept of the "real world"[10] being justified only insofar as it refers to the mental connections that weave the multitude of sense impressions into some connected net. Sense impressions are "conditioned by an 'objective' and by a 'subjective' factor."[11] Similarly, reality itself is a relation between what is outside us and inside us. "The real world is not given to us, but put to us (by way of a riddle)."[12]

Since the world as dealt with by a scientist is more complex than was allowed for in the current philosophies, Einstein thought that the way to escape illusion was by avoiding being a captive of any one school of philosophy. He would take from any system the portions he found useful. Such a scientist, he realized "therefore must appear to the systematic epistemologist as a type of unscrupulous opportunist: he appears as a *realist* insofar as he seeks to describe the world independent of the acts of perception; as *idealist* insofar as he looks upon the concepts and theories as the free inventions of the human spirit (not logically derivable from what is empirically given); as *positivist* insofar as he considers his concepts and theories justified *only* to the extent to which they furnish a logical representation of relations among sensory experiences. He may even appear as *Platonist* or *Pythagorean* insofar as he considers the viewpoint of logical simplicity as an indispensable and effective tool of his research."[13]

But what justifies this "free play with concepts"? There is only one justification: that it can result, perhaps after much labor, in a thought structure which gives us the testable realization of having achieved meaningful order over a large range of sense experiences that would otherwise seem separate and unconnected. In the important essay "Physics and Reality,"[14] which covers much of the same ground as the early pages of the "Autobiographical Notes," Einstein makes the same point with this fine image: "By means of such concepts and mental relations between them we are able to orient ourselves in the labyrinth of sense impressions."[15]

This important process is described by Einstein in a condensed paragraph of the "Autobiographical Notes." "Imagine," he says, "on one side the totality of sense experiences," such as the observation that the needle on a meter is seen to deflect. On the other side, he puts the "totality of concepts and propositions which are laid down in books," which comprises the distilled products of past progress such as the concepts of force or momentum, propositions or axioms that make use of such concepts (for example, the law of conservation of momentum), and more generally, any concepts of ordinary thinking (for example, "black" and "raven"). Investigating the relations that exist among the concepts and propositions is "the business of logical thinking," which is carried out along the "firmly laid-down rules" of logic. The rules of logic, like the concepts themselves, are of course not God-given but are the "creation of humans." However, once they are agreed upon and are part of a widely held convention—the rules of syllogism, for example—they tell us with (only seemingly) inescapable finality that *if* all ravens are black and a particular bird is a raven, then the bird is black. They allow us to deduce from the law of conservation of momentum that in a closed system containing only a neutron and proton, the momentum gained by one is accompanied by the loss experienced by the other. Without the use of logic to draw conclusions, no disciplined thinking, and hence no science, could exist.

But all such conclusions, Einstein warns, are empty of useful "meaning" or "content" until there is some definition by which the particular concept (e.g., "raven" or "neutron") is correlated with actual instances of the concept which have consequences in the world of experience rather than in the world of words and logical rules. Necessary though the correlation or connection between concepts and sense experience is, Einstein warns that it is "not itself of a logical nature." It is an act in which, Einstein holds, "intuition" is one guide, even if not an infallible one. Without it, one could not be led to the assertion that a particular bird, despite some differences in its exact size or degree of blackness from all other birds, does belong to the species raven; or that the start of a particular track, visible in the cloud chamber, is the place where a neutron has struck a proton.

One might wish that Einstein had used a notion more firm than the dangerous-sounding one of "intuition." But he saw no other way. He rejected the use of the word *abstraction* to characterize the transition from observation to concept, e.g., from individual black birds to the idea of "raven." He rejected it precisely because, he said, "I do not consider it justifiable to veil the logical independence of the concept from the sense experiences" (whereas the use of the term *abstraction* or *induction* might make it seem as if there *were* a logical dependence).

The danger is evidently that delusion or fantasy can and does make similar use of the elements of thinking: and since there are no hard, utterly reliable connections between the concepts, propositions, and experience, one cannot know with absolute certainty whether one has escaped the trap of false conclusion. That is why it was thought for so long that observations proved the earth was fixed and the sun went around the earth; that time had a universal meaning, the same for all moving observers; and that Euclidean geometry is the only one that has a place in the physical world. But this is just where Einstein's view is most helpful: Only those who think they *can* play freely with concepts can pull themselves out of such error. His message is even more liberal: The concepts themselves, in our thoughts and verbal expressions, are, "when viewed logically, the free creation of thought which cannot inductively be gained from sense experience." We must be continually aware that it is not necessity but habit which leads us to identify certain concepts (for example, "bread") with corresponding sense experience (feel, smell, taste, satisfaction); for, since this works well enough most of the time, "we do not become conscious of the gulf—logically unbridgeable—which separates the world of sense experience from the world of concepts and propositions." Einstein is perhaps so insistent on the point because he had to discover it the hard way: as a young man, he had to overcome the accepted meanings of such concepts as space, time, simultaneity, energy, etc., and to propose redefinitions that reshaped all our physics, and hence our very concept of reality itself. One might well add here that Einstein demanded the same freedom to challenge orthodoxy outside science. Thus, as a boy he rejected the malignant militarism he saw entrenched in the life of his native country.

Once a conceptual structure has tentatively been erected, how can one check whether it is scientifically "true"? It depends on how nearly the aim of making the system deal with a large amount of diverse sense experience has been achieved, and how economical or parsimonious the introduction of separate basic concepts or axioms into the system has been. Einstein doubted a physical theory, and would say that it failed to "go to the heart of the matter," if it had to be jerry-built with the aid of ad hoc hypotheses, each specially introduced to produce greater agreement between theory and experience (experiment). He also was rarely convinced by theories that dealt with only a small part of the range of physical phenomena, applicable only here or there under special circumstances. In this view, a really good theory, one that has high scientific "truth" value, is considered to be correct not merely when it does not harbor any logical contradictions, but when it allows a close check on the correspondence between the predictions of the theory and a large range of possible experimental experiences. He summarized all this in the following way: "One comes nearer to the most superior scientific goal, to embrace a maximum of experimental content through logical deduction from a minimum of hypotheses. . . . One must allow the theoretician his imagination, for there is no other possible way for reaching the goal. In any case, it is not an aimless imagination but a search for the logically simplest possibilities and their consequences."

This search may take "years of groping in the dark"; hence, the ability to hold onto a problem for a long time, and not to be destroyed by repeated failure, is necessary for any serious researcher. As Einstein once said, "Now I know why there are so many people who love chopping wood. In that activity one immediately sees the results." But for him, the goal of "embracing a maximum of experimental content ... with a minimum of hypotheses" meant nothing less than endless devotion to the simplification and unification of our world picture, for example, by producing fusions in hitherto separate fundamental concepts such as space and time, mass and energy, gravitation and inertial mass, electric and magnetic fields, and inertial and accelerating systems.

KEEPING ALIVE THE SENSE OF WONDER

Embedded in Einstein's views on how to think scientifically about the deep problems, there is an engaging passage in the "Autobiographical Notes" in which Einstein speaks of the importance of the sense of marvel, of deep curiosity, of "wonder," such as his two experiences, when, at the age of four or five, he was shown a magnetic compass by his father, and when, at the age of twelve, a book on Euclidean geometry came into his hands. A person's thought-world develops in part by the mastering of certain new experiences which were so inexplicable, in terms of the previous stage of development, that a sense of wonder or enchantment was aroused. As we learn more, both through science and other approaches, we progressively find that the world around us, as it becomes more rational, also becomes more "disenchanted." But Einstein repeatedly insisted in other writings that there is a limit to this progressive disenchantment, and even the best scientist must not be so insensitive or falsely proud as to forget it. For, as Einstein said in a famous paragraph: "It is a fact that the totality of sense experiences is so constituted as to permit putting them in order by means of thinking— a fact which can only leave us astonished, but which we shall never comprehend. One can say: The eternally incomprehensible thing about the world is its comprehensibility."

He went on: "In speaking here of 'comprehensibility,' the expression is used in its most modest sense. It implies: the production of some sort of order among sense impressions, this order being produced by the creation of general concepts, by relations among these concepts, and by relations of some kind between the concepts and sense experience. It is in this sense that the world of our sense experiences is comprehensible. The fact that it is comprehensible is a wonder."[16]

That wonder [*Wunder*], that sense of awe, can only grow stronger, Einstein implied, the more successfully our scientific thoughts find order to exist among the separate phenomena of nature. This success aroused in him a "deep conviction of the rationality of the universe." To this conviction he gave the name "cosmic religious feeling," and he saw it as the "strongest and noblest motive for scientific research."[17] Indeed, "The most beautiful experience we can have is the mysterious. It is the fundamental emotion which stands at the cradle of true art and true science. Whoever does not know it and can no longer wonder, no longer marvel, is as good as dead. . . ."[18]

After the publication of such sentiments, Einstein received a worried letter from one of his oldest and best friends, Maurice Solovine. They had met in Bern in 1902 when Einstein was twenty-three years old, and they became close friends. Solovine was then a young philosophy student at the University of Bern, to which he had come from Romania, and, together with Conrad Habicht, who was also a student at the university, they banded together to meet regularly to read and discuss works in science and philosophy. With high irony they called themselves the "Olympia Academy." Their "dinners" were no banquets: They all lived on the edge of poverty, and Solovine tells us that their idea of a special dinner was two hard-boiled eggs each. But the talk was that much better, as they discussed works by Ernst Mach, J. S. Mill, David Hume, Plato, Henri Poincaré, Karl Pearson, Spinoza, Hermann Helmholtz, Ampère—and also those of Sophocles, Racine, and Dickens. Many of Einstein's epistemological ideas might be traced back to these discussions.

Now, half a century later, Maurice Solovine was concerned. He asked Einstein how there could be a puzzle about the comprehensibility of our world. For us it is simply an undeniable necessity, which lies in our very nature. No doubt Solovine was bothered that Einstein's remarks seemed to allow into science, that most rational activity of mankind, a function for the human mind which is not rational in the sense of being coldly logical. But Einstein rejected as a "malady"[19] the kind of accusation that implied that he had become "metaphysical."

Instead, he saw the opportunity of using *all* one's faculties and skills to do science as a sign of strength rather than of weakness.

Certainly, he did not propose to abandon rationality, nor to guess where one must puzzle things out in a careful, logical way. But he saw that there is, and has to be, a role for those other elements of thinking which, properly used, can help scientific thought. Specifically, this could become necessary at two points in Einstein's scheme. One is the courageous use of an intuitive feeling for nature *when there is simply no other guide at all*—as when one has tentatively to propose an axiom that by definition is unproved (as Einstein did at the start of the first paper on relativity, where he simply proposed the principle of relativity and the principle of constancy of light velocity); or when one decides which sense experiences to select in order to make an operational definition of a concept. The other point is the sense of wonder at being able to discern something of the grand design of the world, a feeling that motivates and sustains many a scientist.

Einstein's reply (in his letter of 30 March 1952) to Solovine addresses this second point.

> You find it remarkable that the comprehensibility of the world (insofar as we are justified to speak of such a comprehensibility) seems to me a wonder or eternal secret. Now, *a priori*, one should, after all, expect a chaotic world that is in no way graspable through thinking. One could (even *should*) expect that the world turns out to be lawful only insofar as we make an ordering intervention. It would be a kind of ordering like putting into alphabetic order the words of a language. On the other hand, the kind of order which, for example, was created through [the discovery of] Newton's theory of gravitation is of quite a different character. Even if the axioms of the theory are put forward by human agents, the success of such an enterprise does suppose a high degree of order in the objective world, which one had no justification whatever to expect *a priori*. Here lies the sense of "wonder" which increases even more with the development of our knowledge.
>
> And here lies the weak point for the positivists and the professional atheists, who are feeling happy through the consciousness of having successfully made the world not only God-free, but even "wonder-free." The nice thing is that we must be content with the acknowledgment of the "wonder," without there being a legitimate

way beyond it. I feel I must add this explicitly, so you wouldn't think that I—weakened by age—have become a victim of the clergy.

In one of Einstein's other letters to Maurice Solovine, Einstein goes over some of these questions—but this time with the aid of a diagram, as befits a person who prefers to think visually.[20] In this and all these writings, Einstein asks his reader to take the business of making progress in science into one's own hands; to insist on thinking one's own thoughts even if they are not blessed by consent from the crowd; to rebel against the presumed inevitability or orthodoxy of ideas that do not meet the test of an original mind; and to live and think in all segments of our rich world—at the level of everyday experience, the level of scientific reasoning, and the level of deeply felt wonder.

N O T E S

Chapter 1

1. George E. Brown, Jr., quoted in *Science* 260 (1993): 735. I have also analyzed aspects of the antiscience movement in Gerald Holton, *Science and Anti-Science* (Cambridge, Mass.: Harvard University Press, 1993), chapter 6.
2. Leszek Kolakowski, *Modernity on Endless Trial* (Chicago: University of Chicago Press, 1900), p. 4.
3. John Henry Barrows, ed., *The World's Parliament of Religions*, vol. 2 (Chicago: The Parliament Publication Co., 1893), 978–81.
4. W. Ostwald, *Monism as the Goal of Civilization* (Hamburg: International Committee of Monism, 1913), p. 37.
   The section that follows is an abstract of much of "The Controversy over the End of Science," Chapter 5 in Holton *Science and Anti-Science*.
5. H. Stuart Hughes, *Oswald Spengler: A Critical Estimate* (New York: Charles Scribner's Sons, 1952).
6. Rudolf Carnap, Hans Hahn, and Otto Neurath, *Wissenschaftliche Weltauffassung: Der Wiener Kreis* (Vienna: Artur Wolf Verlag, 1929); for an English translation see Otto Neurath, *Empiricism and Sociology* (Dordrecht: Reidel, 1973). The page references are to the English translation; I have made occasional corrections in the translation, as necessary.
7. He allowed a simpler title, *Positivism: A Study in Human Understanding*, for the English translation (Cambridge, Mass.: Harvard University Press, 1951).
8. The word *control*, used in the English edition, has been corrected to *regulate*, which corresponds to the German edition.
9. Seriously mistranslated into English as *Civilization and Its Discontents* (New York: W. W. Norton, 1961).
10. (New York: Simon & Schuster, 1982.) In fairness one must add that Wade since then has embraced a much more balanced view. A particularly

eloquent commentary (N. Wade, *The New York Times Magazine*, 23 July 1995) contains this passage:

> Unlike the useless versifying and arbitrary punishments that filled my school days, science is an ordered realm of theory, laws and facts, where mysteries are resolved and reason prevails, though often not without struggle. Like a slowly spreading pool of light, science now comprehends much of the natural world and is fast reaching to describe the central verities of living things. This achievement bespeaks a degree of understanding that the thinkers of past centuries would have deemed beyond price. An equation that doesn't reflect the hard-won ability to comprehend much of the natural world, and rates science no more necessary than Latin, would befit the Dark Ages.
>
>   Besides which, science is more than just content. It is a rational process in a largely irrational world. It generates knowledge of intrinsic value and of equal and binding value to all races and cultures. Its harvest has never been so rich as now. It is also abstract and sometimes a little difficult. For children who don't learn its central concepts, its doors may stay forever closed.

11. *Science* 26 (1993): 1203.
12. As reported in the *Washington Post*, 20 March 1992.
13. *Nature* 367 (6 January 1994): 6. Unlike most scientific journals in the United States, *Nature* has been alert to the likely damage of the imbalance in reporting. See for example John Maddox's editorial of 17 March 1994 in *Nature* 368 (17 March 1994): 185. It is noteworthy that another among the few who have spoken out against the growing tide of easy condemnation is also a trained science journalist, Barbara J. Culliton, in her essay. "The Wrong Way to Handle Fraud in Science," *Cosmos, 1994*, pp. 34–35.

    For an argument on the costs to science that may result from the excesses of distrust in science, see Steven Shapin, "Truth, Honesty, and Authority of Science," in the National Academy of Sciences report *Society's Choices: Social and Ethical Decision Making in Biomedicine* (Washington, D.C.: National Academy Press, 1994).

14. M. F. Perutz, "The Pioneer Defended," review of *The Private Science of Louis Pasteur,* in *The New York Review of Books*, December 21, 1995.
15. The data were kindly furnished to me by Donald A. B. Lindberg, director, National Library of Medicine. These cases are quite different from the laudable practice of scientists publishing correction notices when they find it necessary to draw attention to their own unintended errors. Eugene Garfield, "How to Avoid Spreading Error," *The Scientist* 1, no. 9 (1987), reports that "of the 10 million journal items indexed in the *SCI* [Science Citation Index] since its inception, over 50,000 were coded as explicit corrections. . . . These vary from corrections of simple typographical errors to retractions of and outright apologies for 'bad' data or data that cannot be verified." This indicates a rate of 0.5 percent for such volunteered corrections of errors.

    In addition, the Office of Research Integrity of the U.S. Public

Health Service recently announced that looking back, it has found a total of 14 researchers guilty of some form of scientific misconduct—out of about 55,000 researchers receiving PHS support per year. (Private communication of 20 July 1993 from Lyle W. Bivens, acting director, ORI.) The cases involved work that ranged over a considerable period; for example, one of them began in 1977. To get a sense of the low yield of the allegations, and the pedestrian rather than sensational nature of most of the cases, see Office of Research Integrity, *Biennial Reports*, U.S. Dept. of Health and Human Services, and the quarterly *O.R.I. Newsletter*.

To glimpse the enormous complexity, cost, and labor as well as the fragility of the process of adjudicating allegations of scientific misconduct, see for example the 63-page document, obtainable from the U.S. Department of Health and Human Services, entitled "Departmental Appeals Board. Research Integrity Adjudications Panel. Subject: Dr. Rameshwar K. Sharma, Docket No. A-93–50, Decision No. 1431, Date: August 6, 1993."

16. For a scholarly and evenhanded treatment of the spectrum of the varied interests of sociologists of science, see Harriet Zuckerman, "The Sociology of Science," in *Handbook of Sociology*, ed. Neil J. Smelser (Beverly Hills, Calif.: Sage Publications, 1988), pp. 511–74.

17. For a thoughtful analysis, see John R. Searle, "Rationalism and Realism, What Is at Stake?," *Daedalus* 122, no. 4 (1993): 55–83. A recent book that aims to answer the various types of critics is Paul R. Gross and Norman Levitt, *Higher Superstition: The Academic Left and Its Quarrels with Science* (Baltimore, Md.: The Johns Hopkins Press, 1994). It is also useful for its extensive bibliography. Another stimulating resource is Frank B. Farrell's *Subjectivity, Realism and Postmodernism* (New York: Cambridge University Press, 1994).

For the "Unabomber's" manifesto, see *Washington Post*, 19 September 1995, supplement. See also *Chronicle of Higher Education*, 11 August 1995, p. A16.

18. Isaiah Berlin, *The Crooked Timber of Humanity, Chapters in the History of Ideas* (New York: Random House, 1992).

19. Alan Beyerchen, *Scientists under Hitler: Politics and the Physics Community in the Third Reich* (New Haven, Conn.: Yale University Press, 1977).

20. Hermann Rauschning, *Gespräche mit Hitler* (New York: Europa Verlag, 1940), p. 210. Mussolini expressed himself similarly.

21. "Politics and the World Itself," *Kettering Review* (Summer 1992): 9–11. His essay was also printed on 1 March 1992 in *The New York Times* as Havel's OpEd, entitled "The End of the Modern Era."

22. Loren R. Graham, *The Ghost of the Executed Engineer: Technology and the Fall of the Soviet Union* (Cambridge, Mass.: Harvard University Press, 1993).

23. Reprinted in *Václav Havel, or Living in the Truth*, ed. Jan Vladislav (London: Faber & Faber, 1987), pp. 138–39. The passage was written in 1984.

24. See note 21. On 4 July 1994 Havel repeated at length much of his previous argument, in the service of explaining the present "state of mind [which] is called postmodernism," and the "crisis" to which science has led mankind. The only new part of his speech (published as an OpEd, 8 July 1994, *The New York Times*) is that our "lost integrity" might paradoxically be renewed by "a science that is new, postmodern," such as the "anthropic cosmological principle" and "the Gaia hypothesis."

   This was too much for Nicholas Wade, who wrote a devastating attack on Havel's essay (N. Wade, "Method and Madness: A Fable for Fleas," *The New York Times Magazine*, 14 August 1994, p. 18), ending with: "A view of the world built on the anthropic principle and the Gaia hypothesis would not be post-modern science but rather a throwback to the numerology and astrology from which the era of rationalism has still not fully rescued us. . . . To subvert rationalism into mysticism would be a cure more pernicious than the disease."

   The seduction of being counted among the postmoderns has apparently attracted even a handful of scientists; the chief example given is their postmodernist interest in "the limits of science." However, the lively discussion of that topic began in the 1870s, led by Emile Dubois-Reymond, and it also preoccupied the logical positivists. For other examples of this old problem, see *Limits of Scientific Inquiry*, ed. G. Holton and R. S. Morison (New York: W. W. Norton, 1978).

25. Published in September 1992 in the *American Journal of Physics* 60 no. 9, (1992): 779–81.

26. Tape recording of the session (12 February 1993) obtainable from the American Association for the Advancement of Science. George Brown's own opening remarks were also distributed as a press release by his Washington, D.C. office.

27. At the 12 February 1993 American Association for the Advancement of Science annual meeting.

28. George E. Brown, "New Ways of Looking at U.S. Science and Technology," *Physics Today* 47 (1994): 32.

   In a talk on "The Roles and Responsibilities of Science in Post-modern Culture" (20 February 1994, at another annual meeting of the American Association for the Advancement of Science), Mr. Brown remarked: "Let me begin by suggesting that the term 'post-modern culture,' sometimes used to describe the present era, is a rubric that comes from the arts and architecture where it had identifiable meaning. To my mind, if the term post-modern is used as a definitional period for policy, politics, or for economic eras, it leads to confusion; and it will not help us to define a point of departure for our discussion here. I hope today's discourse does not get side-tracked on a tedious dissection of post-modernism. I should note, however, that the editorial that appeared in the *New York Times* two years ago entitled 'The End of the Modern Era' by Czech philosopher and playwright Václav Havel, contained several points to which I agree, and have included in previous talks. Although Havel comes to the terms modernism and post-modernism from his artistic and philosophical orientation, I do not subscribe to those labels, in large part because I do not fully understand his use of them."

Similarly, Mr. Brown is one of the few policy makers who has protested Senator Barbara Mikulski's edict that federal funding for basic, "curiosity-driven" research be cut back in favor of supposedly quick-payoff "strategic research."

29. See especially Harvey Brooks, "Research Universities and the Social Contract for Science," in *Empowering Technology: Implementing a U.S. Strategy*, ed. Lewis Branscomb (Cambridge, Mass.: MIT Press, 1993). Brooks has all along been one of the most prescient and observant authors on the place of science in our culture. See for example his essay, "Can Science Survive in the Modern Age?," *Science* 174 (1971): 21–30.

30. E.g., in Don K. Price, "Purists and Politicians," *Science* 163 (1969): 25–31.

Chapter 2

1. T. S. Eliot, *Notes Towards the Definition of Culture* (London: Faber & Faber, 1948).

2. One is reminded of the diagnosis C. P. Snow offered in his provocative book *Science and Government* (Cambridge, Mass.: Harvard University Press, 1961) for the reason why some scientists so single-mindedly stuck to a narrow decision or were satisfied with a narrow range of investigations. It was their success in one particular field or with the operation of one particular apparatus. Snow dubbed these persons "gadgeteers."

3. We also sorely need to give our young scientists more broad humanistic studies—and if I have not dwelled on this it is because, in principle, this can be done with existing programs and facilities, for the tools of study in the humanities, unlike the tools in science, are still to a large degree in touch with our ordinary sensibilities.

4. Edwin A. Burtt, *The Metaphysical Foundations of Modern Physical Science* (New York: Harcourt, Brace & Co., 1927; Atlantic Highlands, N.J.: Humanities Press, 1995), pp. 236–37.

5. Arthur Koestler, *The Sleepwalkers* (London: Hutchinson, 1959), p. 518.

6. Ibid., pp. 529–31.

7. Ibid., pp. 541–42.

8. Lionel Trilling, *Mind in the Modern World: The 1972 Jefferson Lecture in the Humanities* (New York: Viking Press, 1972), pp. 13–14.

Chapter 3

1. See, e.g., Owen Gingerich, " 'Crisis' versus Aesthetic in the Copernican Revolution," in *Copernicus: Yesterday and Today*, ed. Arthur Beer and K. Aa. Strand (Oxford: Pergamon Press, 1975), pp. 85–93.
2. N. Copernicus, *De revolutionibus orbium coelestium* (Nuremberg: 1543), f. iii (v).
3. See Carl Seelig, *Albert Einstein* (Zurich: Europa Verlag, 1954), p. 195.
4. H. A. Lorentz, *The Einstein Theory of Relativity* (New York: Brentano's, 1920), pp. 23–24.
5. H. C. Oersted, "On the Spirit and Study of Universal Natural Philosophy," in *The Soul of Nature*, trans. Leonora and Joanna B. Horner (London, 1852; reprint, London, 1966), p. 450.
6. Galileo Galilei, *Dialogues Concerning Two New Sciences*, trans. Henry Crew and Alfonso de Salvio (New York: Macmillan, 1914), p. 179.
7. Alexandre Koyré, *Galileo Studies*, trans. John Mepham (Atlantic Highlands, N.J.: Humanities Press, 1978).
8. Galileo, p. 170.
9. Galileo, p. 153.
10. See Steven Shapin and Simon Schaffer, *Leviathan and the Air Pump: Hobbes, Boyle and the Experimental Life* (Princeton, N.J.: Princeton University Press, 1989).
11. H. C. Oersted, "Experiments on the Effect of a Current of Electricity on the Magnetic Needle," *Annals of Philosophy* 16 (1820): 273–76. This is the English translation of Oersted's pamphlet, *Experimenta circa effectum conflictus electrici in acum magneticam.*
12. Oersted, pp. 273–74.
13. For some of these estimates, see G. L'E. Turner, "The Microscope as a Technical Frontier in Science," in *Historical Aspects of Microscopy*, ed. S. Bradbury and G. L'E. Turner (Cambridge, England: W. Heffer & Sons, 1967), pp. 175–97; on the astronomical angular measures, see H. T. Pledge, *Science Since 1500* (New York: Harper, 1959), p. 291. On the pp̄ ratios, see G. Gabrielse, et al., *Physical Review Letters* 74 (1995): 3544–47.
14. See James Bryant Conant, ed., *Harvard Case Histories in Experimental Science. Case 1: Robert Boyle's Experiments in Pneumatics* (Cambridge, Mass.: Harvard University Press, 1967), p. 62.
15. Dong-Won Kim, "The Emergence of the Cavendish School: An Early History of the Cavendish Laboratory, 1871–1900" (Ph.D. diss. Harvard University, 1991).
16. Sir James Chadwick, "Some Personal Notes on the Search for the Neutron," delivered at the Tenth International Congress of History of Science at Cornell University, Ithaca, New York, in 1962; reprinted in *The Project*

Physics Course Reader (New York and Toronto: Holt, Reinhart & Winston, 1971), unit 6, p. 28.

17. E. Fermi, E. Amaldi, B. Pontecorvo, F. Rasetti, and E. Segrè, *Ricerca Scientifica* 5, no. 2 (1934): 282–283; translated and reprinted with the title "Influence of Hydrogenous Substances on the Radioactivity Produced with Neutrons—I," in Enrico Fermi, *Collected Papers*, vol. 1 (Chicago: University of Chicago Press, 1962), pp. 761–62.

18. F. Abe, et al., "Limit on the Top-Quark Mass from Proton-Antiproton Collisions at $\sqrt{s} = 1.8$ TeV," *Physical Review D* 45 (1992): 3921; F. Abe, et al., *Physical Review Letters*, 74 p. 2626 (1995); and S. Abachi, et al., *Ibid*, p. 2632.

19. See Gerald Holton, "Subelectrons, Presuppositions, and the Millikan-Ehrenhaft Dispute," in *The Scientific Imagination* (New York: Cambridge University Press, 1978), pp. 25–83.

20. Robert A. Millikan, "A New Modification of the Cloud Method of Determining the Elementary Electric Charge and the Most Probable Value of that Charge," *Philosophical Magazine* 19 (1910): 209–228.

21. Millikan, p. 220.

22. Millikan, pp. 221–23.

23. Philip R. Bevington and D. Keith Robinson, *Data Reduction and Error Analysis for the Physical Sciences*, 2d ed. (New York: McGraw-Hill, 1992).

24. E. Bright Wilson, Jr., *An Introduction to Scientific Research* (New York: McGraw-Hill, 1952; reprint, New York: Dover Publications Inc., 1990).

   See Theodore M. Porter, *The Rise of Statistical Thinking 1820–1900* (Princeton, N.J.: Princeton University Press, 1986) for a guide to the development of the concern with statistics in science.

25. Peter Galison, *How Experiments End* (Chicago: University of Chicago Press, 1987).

26. Ibid., p. 183.

27. Ibid., pp. 193–194.

28. F. J. Hasert et al., "Observation of Neutrino-like Interactions without Muon or Electron in the Gargamelle Neutrino Experiment," *Physics Letters B* 46 (1973): 138–140.

29. See, for instance, the letter from Oscar Hernandez in *Science* 258 (2 October 1992): 13. On federal courts ordering the "sharing" of confidential data, see *Science* 261 (July 1993): 284–285.

30. After accepting the existence of neutral currents, one of the chief investigators wrote: "Three pieces of evidence now in hand point to the distinct possibility that a [muon]less signal of order 10% is showing up in the data. At present I don't see how to make these effects go away." Galison, p. 235.

31. See Faye Flam, "Big Physics Provokes a Backlash," *Science* 257 (September 1992): 1468. The tendency toward a new organization of scientific

work is described in John Ziman, *Of One Mind: The Collectivization of Science* (New York: American Institute of Physics Press, 1995).

32. P. W. Bridgman, *Reflections of a Physicist* (New York: Philosophical Library, 1955), p. 535. The passage from this source reads: "The scientific method, as far as it is a method, is nothing more than doing one's damnedest with one's mind, no holds barred."

33. P. W. Bridgman in *Collected Experimental Papers*, ed. Harvey Brooks, Francis Birch, Gerald Holton, and William Paul (Cambridge, Mass.: Harvard University Press, 1964).

34. P.W. Bridgman, " 'Manifesto' by a Scientist," *Science* 89 (1939): 179.

Chapter 4

1. From an unpublished manuscript, quoted in G. Holton, "Notes Toward the Psychobiographical Study of Scientific Genius," in Y. Elkana, ed., *The Interaction Between Science and Philosophy* (New York: Humanities Press, 1975), pp. 370–71.

2. Howard M. Georgi, "Grand Unified Theories," in *The New Physics*, ed. Paul Davies (Cambridge, England: Cambridge University Press, 1989), p. 435.

3. Thomas Young, "Outline of Experiments and Inquiries Respecting Sound and Light," in *Miscellaneous Works of the Late Thomas Young*, ed. George Peacock (London: John Murray, 1855), pp. 80–81.

4. Erwin Panofsky, "Galileo as a Critic of the Arts: Aesthetic Attitude and Scientific Thought," *Isis* 47 (1956): 3–15.

5. Quoted in Panofsky, p. 5.

6. Panofsky, p. 7.

7. Panofsky, p. 10.

8. Quoted in Panofsky, p. 13.

9. Further Reading for Chapter 4.

    Edgerton, Samuel Y. Jr. "Galileo, Florentine 'Disegno,' and the 'Strange Spottednesse' of the Moon." *Art Journal* (Fall 1984): 225–32; Kemp, Martin, *The Science of Art*. New Haven: Yale University Press, 1990; and essays on this case by M. Nicolson, I. B. Cohen, and M. Biagioli.

    Galison, Peter. *How Experiments End*. Chicago, Ill.: University of Chicago Press, 1987.

    Hadamard, Jacques. *The Psychology of Invention in the Mathematical Field*. New York: Dover Publications, 1954.

    Holton, Gerald. "Metaphors in Science and Education," in *The Advancement of Science, and Its Burdens*. New York: Cambridge University Press, 1986, pp. 229–52.

    Holton, Gerald. "The Thematic Imagination in Science," pp. 31–52, and "Thematic and Stylistic Interdependence," pp. 75–98, in *Thematic Origins of Scientific Thought: Kepler to Einstein*, rev. ed. Cambridge, Mass.: Harvard University Press, 1988.

Miller, Arthur I. *Imagery in Scientific Thought: Creating 20th-Century Physics*. Boston: Birkhäuser, 1984.

Chapter 5

1. The *Guide to the History of Science*, published by the History of Science Society, lists thirty-seven subsets of subject interests of its members, from Greek and Roman antiquity to contemporary science, technology, and medicine.
2. P. W. Medawar, *The Art of the Soluble* (London: Methuen & Co., Ltd., 1967), p. 7.
3. Ibid, pp. 151, 155.
4. In this section, I refer to certain details in case studies on relativity theory to be found in four of my books: *Thematic Origins of Scientific Thought: Kepler to Einstein* (Cambridge, Mass.: Harvard University Press, 1988); *The Scientific Imagination: Case Studies* (New York: Cambridge University Press, 1978); *The Advancement of Science, and Its Burdens: The Jefferson Lecture and Other Essays* (New York: Cambridge University Press, 1986), and *Science and Anti-Science* (Cambridge, Mass.: Harvard University Press, 1993).
5. Richard P. Feynman, "The Development of the Space-Time View of Quantum Electrodynamics," in *Les Prix Nobel en 1965* (Stockholm: Imprimerie Royal P. A. Norstedt & Sönner, 1966).
6. A detailed discussion of this conception, and studies illustrating its influence in specific cases, are given in the three books mentioned in note 4, above and in Part Two.
7. Einstein to Born, 3 March 1947. Translated from Albert Einstein and Hedwig and Max Born, *Briefwechsel 1916–1955* (München: Nymphenburger Verlagshandlung, 1969), p. 215. See also *The Born-Einstein Letters*, trans. Irene Born (New York: Walker and Co., 1971), p. 158.
8. Quoted in Leonard Huxley, *The Life and Letters of Thomas Henry Huxley* (London: MacMillan, 1900).
9. P. A. M. Dirac, "The Early Years of Relativity," in *Albert Einstein: Historical and Cultural Perspectives*, ed. G. Holton and Y. Elkana (Princeton, N.J.: Princeton University Press, 1982), pp. 84–85.

Chapter 6

1. Daniel N. Lapedes, ed., *McGraw-Hill Dictionary of Scientific and Technical Terms*, 2d ed. (New York: McGraw-Hill, 1978), pp. 512–513. As another measure in the continuing, albeit sometimes only ritualistic, reference made in the ongoing research literature to Einstein's publications, Eugene Garfield has found that during the period 1961–1975 the serious scientific journals in toto carried no less than 40 million citations to previously published articles. Of these, 58 cited articles stand out by virtue of having

been published before 1930 *and* cited over 100 times each; and among these 58 enduring classics, ranging from astronomy and physics to bio-medicine and psychology, 4 are Einstein's, See E. Garfield, *Current Contents* 21 (1976): 5–9.

2. In Harry Woolf, ed., *Some Strangeness in the Proportion: A Centennial Symposium to Celebrate the Achievements of Albert Einstein* (Reading, Mass.: Addison-Wesley Publishing Co., Inc., 1980), p. 108.

3. Quoted in Ernst Lechner, *Physikalische Weltbilder* (Leipzig: Theodore Thomas Verlag, 1912), p. 84.

4. Quoted in the *New York Times*, 3, 4, and 5 April 1921.

5. For discussion and documentation, see G. Holton, "Einstein's Search for the '*Weltbild*,' " Chapter 4 in G. Holton, *The Advancement of Science and Its Burdens* (New York: Cambridge University Press, 1986).

6. Quoted from J. J. Thomson, *Reflections and Recollections* (London: G. Bell and Sons, Ltd., 1936), p. 431 (italics in original). See also Philipp Frank, *Einstein: His Life and Times* (New York: Alfred A. Knopf, 1947), p. 190. Frank's book is one of the good sources for documentation on the reception and rejection of Einstein's theories by various religions and philosophic and political systems.

7. These articles, and excerpts from some other publications dealing with the influence of Einstein's work, have been gathered in L. Pearce Williams, ed., *Relativity Theory: Its Origins and Impact on Modern Thought* (New York: John Wiley and Sons, 1968). I am indebted to it for a number of illustrations referred to in this chapter.

8. John Passmore, *A Hundred Years of Philosophy*, rev. ed. (New York: Basic Books, 1966), p. 332.

9. As reported by Paul M. LaPorte, "Cubism and Relativity, with a Letter of Albert Einstein," *Art Journal 25*, no. 3 (1966): 246.

10. Ibid. See also C. H. Waddington, *Behind Appearances: A Study of the Relations between Painting and the Natural Sciences in this Century* (Edinburgh: Edinburgh University Press, 1969; Cambridge, Mass.: MIT Press, 1970), pp. 9–39. At the Jerusalem Symposium in 1979, Meyer Schapiro presented an extensive and devastating critique of the frequently proposed relation between modern physics and modern art.

11. *Contact* 4 (1923): 3. I am indebted to Carol Donley's draft paper, "Einstein, Too, Demands the Muse," for this lead and others in the following paragraphs.

12. *Selected Essays of William Carlos Williams* (New York: Random House, 1954), p. 283.

13. Ibid., p. 340.

14. J.-P. Sartre, "François Mauriac and Freedom," in *Literary and Philosophical Essays* (New York: Criterion Books, 1955), p. 23.

15. Lawrence Durrell, *Balthazar* (New York: E. P. Dutton, 1958), Author's Note, p. 9.

16. Ibid., p. 142.

17. For a good review of details, to which I am indebted, see Alfred M. Bork, "Durrell and Relativity," *Centennial Review* 7 (1963): 191–203.

18. L. Durrell, *A Key to Modern British Poetry* (Norman, Okla.: University of Oklahoma Press, 1952), p. 48.

19. Ibid., pp. 25, 26, 29.

20. All quotations are from William Faulkner, *The Sound and the Fury* (London: Chatto and Windus, 1961), pp. 81–177. I thank Dr. J. M. Johnson for a draft copy of her interesting essay "Albert Einstein and William Faulkner," and have profited from some passages even while differing with others.

21. The chapter is shot through with references to light, light rays, even to travel "down the long and lonely light rays."

22. In *Les Prix Nobel en 1950* (Stockholm: Imprimerie Royale, 1951), p. 71.

23. Jean Piaget, *The Child's Conception of Time* (New York: Ballantine Books, 1971), p. vii.

24. Jean Piaget, *Genetic Epistemology* (New York: Columbia University Press, 1970), p. 69; see also p. 7.

25. Jean Piaget, *Psychology and Epistemology* (New York: Grossman Publishers, 1971), p. 82; see also pp. 10, 110. A similar statement is to be found in Piaget's *Six Psychological Studies* (New York: Vintage Books, 1968), p. 85.

26. For example, Jean Piaget, with Bärbel Inhelder, *The Child's Conception of Space* (New York: W. W. Norton, 1967), pp. 232–233; *The Child's Conception of Time* (London: Routledge and Kegan Paul, 1969), pp. 305–306; *Biology and Knowledge* (Chicago: University of Chicago Press, 1971), pp. 308, 337, 341–342. I wish to express my thanks to Dr. Katherine Sopka for help in tracing these references.

27. All this is quite apart from the role of model or cultural hero that Einstein played in the lives of a great many individuals whom he never met. To this day, his picture can be found in wide circulation and in the most unlikely places, from the T-shirt of a student in high school to the workbench of a cobbler in Shanghai. As befits this period of revisionism, there are also a few—chiefly journalists—who have found ready audiences for a demonized version of this scientist (among many others).

28. Albert Einstein, *Über die spezielle und die allgemeine Relativitätstheorie, gemeinverständlich* (Braunschweig: Vieweg, 1917). It was often translated and to this day is perhaps his most widely known work.

29. Albert Einstein and Leopold Infeld, *The Evolution of Physics* (New York: Simon & Schuster, 1938).

30. Albert Einstein *On the Method of Theoretical Physics* (Oxford: Clarendon Press, 1933).

31. First published in English in 1933, in his *The Modern Theme* (New York: W. W. Norton, 1933).

32. Ibid., pp. 135–136.

Chapter 7

1. George Sarton, *The Study of History of Science* (1936; reprint, New York: Dover, 1957), p. 5.
2. Pierre Duhem, *The Aim and Structure of Physical Theory* (New York: Atheneum, 1962), p. 177.
3. P. A. M. Dirac, *Principles of Quantum Mechanics* (Oxford: Clarendon Press, 1958), p. 312.
4. Victor Weisskopf, "Of Atoms, Mountains, and Stars: A Study in Qualitative Physics," *Science* 187, no. 4177 (1975): 605–12.
5. Robert G. Sachs, "Structure of Matter: A Five-Year Outlook," *Physics Today* 32, no. 12 (December 1979): 27.
6. Usually translated as "On the Method of Theoretical Physics," this was Einstein's Herbert Spencer Lecture, given at Oxford on 10 June 1933. The original manuscript of Einstein's essay was in German and has been published in his collection *Mein Weltbild* (Frankfurt am Main: Ullstein, 1977), pp. 113–19. An English translation appeared as a small booklet by Oxford University Press in 1933; but it left a good deal to be desired, and a different English translation was prepared (by Sonja Bergmann) when Einstein later published a collection of his essays in *Ideas and Opinions* (New York: Crown, 1954). The translation appears on pp. 270–76. In quoting from Einstein's essay and from his other writings, I refer to the pages of the English translations in *Ideas and Opinions*, but I have gone back to the corresponding original German essays and corrected the published English versions as necessary. In this connection, I wish to acknowledge with thanks the permission of the estate of Albert Einstein to quote from Einstein's writings.
7. Philipp Frank, *Einstein: His Life and Times* (New York: Knopf, 1947), p. 217.
8. Quoted in Richard K. Gehrenbeck, "C. J. Davisson, L. H. Germer, and the Discovery of Electron Diffraction" (Ph.D. diss., University of Minnesota, 1973), pp. 343–44.
9. After the observation of the bending of starlight passing by the sun, published in November 1919, Einstein amended this sentence for the edition printed in 1920: Now there remained only one consequence drawn from the theory which had not been observed (the red shift of spectral lines); but again, he added, "I do not doubt at all that this consequence of the theory will also find its confirmation soon." Albert Einstein, *Über die spezielle und die allgemeine Relativitätstheorie*, 7th ed. (Braunschweig: Vieweg, 1920), p. 70.
10. E.g., Einstein, *Ideas and Opinions*, p. 272. The quotations in the next six paragraphs are from the same source, pp. 273–76.
11. In the same essay, "Physik und Realität," 1936, translated as "Physics and Reality," in Einstein, *Ideas and Opinions*, p. 294.
12. In Pierre Speziali, ed., *Albert Einstein, Michele Besso, Correspondence 1903–1955* (Paris: Hermann, 1972), p. 527.

13. Isaiah Berlin, *Concepts and Categories* (New York: Viking Press, 1979), p. 159.
14. Albert Einstein, in *Albert Einstein: Philosopher-Scientist*, ed. Paul A. Schilpp (Evanston, Ill.: Library of Living Philosophers, 1949), p. 53.
15. Albert Einstein, "Induktion und Deduktion in der Physik," *Berliner Tageblatt* (Supplement), 25 December 1919.
16. Einstein, in *Albert Einstein*, pp. 673–74. See also p. 678: "Categories are necessary as indispensable elements of thinking."
17. A brief survey of thematic analysis and some case studies are provided in Gerald Holton, *Thematic Origins of Scientific Thought: Kepler to Einstein* (Cambridge, Mass.: Harvard University Press, 1988), and in Gerald Holton, *The Scientific Imagination: Case Studies* (New York: Cambridge University Press, 1978). The "Postscript" in *Thematic Origins* discusses the uses of thematic analysis in many fields, by other scholars.
18. Max Planck, "Verhältnis der Theorien zueinander," in *Die Kultur der Gegenwart*, part 3, vol. 1, ed. Paul Hinneberg (Leipzig: B. G. Teubner, 1915), p. 737.
19. Albert Einstein, *Aether und Relativitätstheorie* (Berlin: Julius Springer, 1920), p. 14.
20. J. T. Merz, *A History of European Thought in the Nineteenth Century* (London: William Blackwood & Sons, 1904), I, pp. 251–52.
21. Ludwig Büchner, *Kraft und Stoff: Empirisch-naturphilosophische Studien*, 9th ed. (Leipzig: Theodor Thomas, 1867), p. 89.
22. Ernst Mach, *Die Mechanik in ihrer Entwicklung, historisch-kritisch dargestellt*, 2d ed. (Leipzig: F. A. Brockhaus, 1889), pp. 437–38.
23. Cf. "Aufruf," *Physikalische Zeitschrift*, 13 (1912): 735–36; and Friedrich Herneck, "Albert Einstein und der philosophische Materialismus," *Forschungen und Fortschritte* 32 (1958): 206. For an analysis of the document, and its sources and influence, see "Ernst Mach and the Fortunes of Positivism," Chapter 1 in G. Holton, *Science and Anti-Science* (Cambridge, Mass.: Harvard University Press, 1993).
24. Albert Einstein, "Motiv des Forschens"; a somewhat loose English translation was published in Einstein, *Ideas and Opinions*, pp. 224–27.
25. In *Albert Einstein*, pp. 59–61, 81; emphases in original. In the 1933 Oxford lecture, Einstein raises this problem only gently, and at the end, by saying: "Meanwhile the great stumbling block for a field theory of this kind lies in the conception of the atomic structure of matter and energy. For the theory is fundamentally nonatomic insofar as it operates exclusively with continuous functions of space" (p. 275), unlike classical mechanics, which by introducing as its most important element the material point does justice to an atomic structure of matter.
26. Einstein, *Ideas and Opinions*, p. 272. The phrase was translated in the first English version of the 1933 lecture delivered by Einstein as "the adequate representation of a single datum of experience."

27. Einstein, *Physik und Realität*, p. 318.
28. The case is quite general. Thus, Kepler's world was constructed of three overlapping thematic structures, two ancient and one new: the universe as theological order, the universe as mathematical harmony, and the universe as physical machine. Newton's scientific world picture clearly retained animistic and theological elements. Lorentz's predominantly electromagnetic worldview was really a mixture of Newtonian mechanics as applied to point masses, determining the motion of electrons, and Maxwell's continuous-field physics. Ernest Rutherford, writing to his new protégé, Niels Bohr, on 20 March, 1913, gently scolds him: "Your ideas as to the mode of origin of spectra in hydrogen are very ingenious and seem to work well; but the mixture of Planck's ideas [quantization] with the old mechanics makes it very difficult to form a physical idea of what is the basis of it." In fact, of course, Bohr's progress toward the new quantum mechanics via the correspondence principle was a conscious attempt to find his way stepwise from the classical basis.
29. Einstein, "On the Theory of Relativity," in *Ideas and Opinions*, p. 246.

Chapter 8

1. John Stachel, "Albert Einstein and Mileva Marić: A Collaboration that Failed to Develop," in *Creative Couples in Science*, ed. H. Pycior, N. Slack, P. Abir-Am (New Brunswick, N.J.: Rutgers University Press, 1995). Parts of my presentation are indebted to his work.
2. Some of the relevant work of sociologists and psychologists on this subject has been summarized in Gerald Holton, Chapter 7 in "On the Psychology of Scientists, and Their Social Concerns," in *The Scientific Imagination: Case Studies* (New York: Cambridge University Press, 1978).
3. Albert Einstein, "Autobiographical Notes," pp. 3–4, in *Albert Einstein: Philosopher-Scientist*, ed. Paul A. Schilpp (La Salle, Ill.: Open Court, 1949).
4. Some journalists have recently sought for the genetic origin of Eduard's mental condition not in Mileva's family but in Albert himself. Their source is a dubious theory of "creativity" promoted by Anthony Storr (*The Dynamics of Creation* [New York: Athenaeum, 1972]), who begins by announcing that "creative activity [is] a peculiarly apt way for the schizoid individual to express himself" (p. 57). Then he selects his prime target as follows: "Einstein provides the supreme example of how schizoid detachment can be put to creative use," although generously allowing that "he retained sufficient contact with reality for his thought to be scientifically viable" (pp. 61, 63). But Storr warns that the price to be paid for the "detachment" needed for making "a new model of the universe" is severe: "Such detachment can only be achieved by the person with a predominantly schizoid psychopathology" (p. 67). From there, Dr. Storr moves on to deal similarly with Isaac Newton. The whole argument reads like a parody of ancient

psychiatric work; but it fits with today's trend to construct revisionist accounts of major innovators.

5. John Stachel, "Albert Einstein and Mileva Marić: A Collaboration that Failed to Develop," p. 217.

6. Bibliographical Sources for Chapter 8.

Einstein, Elizabeth Roboz. *Hans Albert Einstein: Reminiscences of His Life and Our Life Together*. Iowa City: University of Iowa, 1991.

Erikson, Erik H. "Psychoanalytic Reflections on Einstein's Centenary," in *Albert Einstein: Historical and Cultural Perspectives*, ed. G. Holton and Y. Elkana. Princeton, N.J.: Princeton University Press, 1982.

Frank, Philipp. *Einstein: His Life and Times*. New York: Alfred A. Knopf, 1947.

Michelmore, Peter. *Einstein: Profile of the Man*. New York: Dodd, Mead & Co., 1962.

Pais, Abraham. *Einstein Lived Here*. New York: Oxford University Press, 1994.

Reiser, Anton. *Albert Einstein*. New York: Albert & Charles Boni, 1930.

Renn, Jürgen, and Robert Schulmann, eds. *Albert Einstein, Mileva Marić: The Love Letters*. Translated by Shawn Smith. Princeton, N.J.: Princeton University Press, 1992.

Stachel, John, ed. *The Collected Papers of Albert Einstein, Vol. I: The Early Years, 1879–1902*. Princeton, N.J.: Princeton University Press, 1987.

Stachel, John. "Einstein and Marić: A Failed Collaboration," in *Creative Couples in Science*, ed. H. Pycior, N. Slack, and P. Abir-Am. New Brunswick, N.J.: Rutgers University Press, 1995.

Trbuhović-Gjurić, Desanka. *Im Schatten Albert Einsteins*. Bern and Stuttgart: Verlag Paul Haupt, 1988. [But see the corrections of the claims, by Pais and Stachel above. The book is a translation from the original in Serbo-Croatian (*U senci Alberta Ajnstajna*, 1969), with many unprovable or plainly contrafactual assertions which appeal to sensationalizing journalistic versions such as Andrea Gabor, *Einstein's Wife* (New York: Viking, 1995).]

Note: The translations from the German originals of the letters are partly based on existing ones, but are largely the author's. I gladly acknowledge permission from the Einstein Estate to quote from the letters.

Chapter 9

1. Quoted from a report by Banesh Hoffmann in *Einstein, the Man and His Achievement*, by G. J. Whitrow (New York: Dover, 1967), p. 75.

2. Paul Arthur Schilpp, ed., *Albert Einstein: Philosopher-Scientist* (Evanston, Ill.: Library of Living Philosophers, Inc., 1949), pp. 683–84.

3. Albert Einstein, "Autobiographical Notes" (in German and English), in Schilpp, pp. 1–95.

4. Ibid., p. 33.

5. Albert Einstein, *Ideas and Opinions*, new translations and revisions by Sonja Bergmann (New York: Crown, 1954), pp. 25–26.

6. Ibid., p. 22.

7. Ibid.

8. Ibid., p. 301.

9. Ibid., p. 20.

10. Ibid., p. 291.

11. Schilpp, p. 673.

12. Ibid., p. 680.

13. Ibid., p. 684.

14. Einstein, *Ideas and Opinions*, p. 290ff.

15. Ibid., p. 291

16. Ibid., p. 292 (but in corrected translation).

17. Ibid., p. 39.

18. Ibid., p. 11.

19. Ibid., p. 24.

20. The analysis is the subject of "Einstein's Model for Constructing a Scientific Theory," chapter 2 in G. Holton, *The Advancement of Science, and Its Burdens: The Jefferson Lecture and Other Essays* (New York: Cambridge University Press, 1986).

ACKNOWLEDGMENTS

I wish to thank Joan Laws for her splendid help in guiding the volume from manuscript to publication, and the original publishers for permission to use parts of my essays in their new form. The following list gives the primary sources on which the chapters of this book were based; in each case, the material was reworked for this publication.

Chapter 1 is based on a paper delivered at the 1993 Sigma Xi Forum, "Ethics, Values, and the Promise of Science," 26 February 1993 (published in the *Forum Proceedings* by Sigma Xi in 1993), and on a lecture given at the University of Vienna on 20 April 1993 (published in *Public Understanding of Science* 2 [1993]).

Chapter 2 is adapted from "Modern Science and the Intellectual Tradition," Chapter 14 in G. Holton, *Thematic Origins of Scientific Thought: Kepler to Einstein* (Cambridge, Mass.: Harvard University Press, 1973).

Chapter 3 is based on the *Miller Lecture in Science and Ethics*, given at the Massachusetts Institute of Technology on 7 April 1990.

Chapter 4 is based on a paper delivered at Spoletoscienza Festival, Spoleto, Italy, 29 June 1991, and on "L'immaginazione nella scienza," Chapter 7 in G. Holton, *La Responsibilitá della Scienza* (Rome: Gius. Laterza & Figli, 1993).

Chapter 5 is based on a lecture in a series given during a lecture tour of universities in China, 1985.

Chapter 6 is based on the introductory chapter, "Einstein and the Shaping of Our Imagination," in *Albert Einstein: Historical and Cultural Perspectives*, ed. Gerald Holton and Yehuda Elkana (Princeton, N.J.: Princeton University Press, 1982).

Chapter 7 is based on the chapter "Toward a Theory of Scientific Progress," in *Progress and Its Discontents*, ed. G. A. Almond, M. Chodorow, and R. H. Pearce (University of California Press, 1982).

Chapter 8 is based on my articles in *Physics Today* 47 (Part 1, August, and Part 2, September 1994).

Chapter 9 uses material that appeared in *The Physics Teacher* 17, no. 3 (March 1979).

I N D E X

5/4/98